“Recommended
at all

★ ★ ★

The Defense of Hill 781 is entertaining and has many valuable tactical lessons to offer to today’s infantry leader. It is recommended reading for leaders at all levels.

Infantry Magazine

McDonough’s battle descriptions are the most accurate I have ever read. Having been through the NTC as a battalion S3 (operations and training officer), I found that this book brings back both bitter and sweet memories. . . . [Lessons learned] are accurate and realistic and would provide an excellent check list for anyone en route to the NTC, particularly battalion commanders.

[McDonough] also presents a vivid picture of the confusion and turmoil during each battle that affect a battalion’s plans and its mission. He discusses the need for continuous operational planning, rehearsals and coordination. The lessons learned apply to every branch and every facet of battalion operations and tactics.

This book will soon be on every professional soldier’s reading list and rightfully so. It is easy to read and one you cannot put down. Get your copy early; they will not be on the shelves for very long.

Frank J. Grand III
Combined Arms Combat Development Activity
Fort Leavenworth, Kansas, in *Military Review*

★ ★ ★

★ ★ ★

While the American way of war may differ from ours, the difference is in style rather than substance. Therefore the lessons learned are equally applicable to the British Army. It is for this reason that *The Defense of Hill 781* is an important book which is strongly recommended for any officer.

British Army Review

The Defense of Hill 781 is more than a well-written tactical treatise on modern warfare. Jim McDonough offers us insight into both the tactical competence and character of command required for decisive victory. The author's description of the chemistry that bonds cohesive units in battle has the ring of truth that combat veterans will recognize. His tactical primer on attaining the collective force of decentralized, dispersed units, synchronized with the common vision of the commander's intent and energized with the freedom to act is an important professional lesson for anyone who would lead troops in battle. If you want a deeper understanding of why the Iraqi Army dissolved in 100 hours when facing a coalition led by U. S. forces trained for the past decades in the harsh, realistic crucible of the National Training Center . . . read this book.

LTG J. W. Woodmansee, USA (Ret.)

★ ★ ★

The Defense of Hill 781

An Allegory of Modern Mechanized Combat

Also by James R. McDonough

PLATOON LEADER
LIMITS OF GLORY

The Defense of Hill 781

An Allegory of Modern Mechanized Combat

James R. McDonough

Foreword
by
Gen. John R. Galvin, USA (Ret.)

★
Presidio

This edition published in 1993

Published by Presidio Press
505 B San Marin Dr., Suite 300
Novato, CA 94945

Library of Congress Cataloging-in-Publication Data

McDonough, James R., 1946–
The defense of Hill 781.

1. Tactics. 2. Military maneuvers.
3. United States. Dept. of the Army.
National Training Center. 4. United States.
Army. I. Title.
U167.M49 1988 355.4'2 87–36026
ISBN 0-89141-475-4

Printed in the United States of America

Contents

Acknowledgments

With appreciation to Hal Winton and Ted Pusey, who gave me the idea for this book, and in memory of Sergeant Masterson who gave his all for the readiness of our Army.

This book is dedicated to America's straight and stalwart soldiers who have trained hard in times of peace to be ready for war, and have thereby kept the peace.

Foreword

One of the best thinkers in today's United Stated Army has written what at first seems to be a lighthearted and simple story. A lieutenant colonel dies (from eating the army's rations) and finds himself in Purgatory, which turns out to be a place in the high desert not far from Barstow, California. Here he must atone for his sins and prove himself worthy of Heaven by leading a battalion of soldiers against a well-armed and hard-bitten enemy.

The place is the National Training Center, the army's premier installation for development of tactical leaders through hands-on application—but of course the place is also Purgatory, where souls pay for the sins of a life that is now past. The weapons are real—the rifles fire bullets, not blanks, and the battle is for keeps. At the same time, there are observers who monitor and critique the actions of the participants. McDonough's imagination takes the reader from the real world to the world of games, and from there, well, to Hell and back—or at least to a very tough Purgatory—and as the book goes along, all of these places become one and the same in a compelling and fascinating departure from worldly reality. The reader becomes the victim of a cordial kidnapping and is led off into circumstances that require what Coleridge called "a willing suspension of disbelief."

Jim McDonough deftly mixes several sets of imagery to stretch the imagination of the reader and move him into a fascinating world of alternating fancy and fact. There are many levels of experience captured here, all flavored with a subtle and understated humor that any soldier, and especially those who have gone through the National Training Center, will find delightful.

What is the meaning of all this? The secret: McDonough's book is a training text, a compilation of lessons learned in this desert school of hard knocks, presented as the story of a single leader's trials and errors as he fights and learns his way to salvation. The real National Training Center, conceived in the late 1970s and built at Camp Irwin, California, where World War II soldiers learned desert fighting, is now a marvel of technological developments where instruments follow every action and every communication to provide the ability to review in detail the clash of military units. The fighters themselves, opposed by an ominously capable high-technology enemy, fire laser beams instead of bullets and reach a level of reality of combat that provides them that most important opportunity—the chance to fight and "die" and rise again to fight another day (or night) and thus to learn what before only real battle and real death could teach. There will never be a way to tell how many lives the National Training Center saved already on the battlefields of Desert Storm, and how many it will save in the future, but whatever the number, this place is providing the most productive, the most useful training any army has ever had. McDonough tells its story, and describes what there is to be learned there, better than anyone has done or probably will do.

McDonough's whimsical hero, Lieutenant Colonel A. Tack Always, is a good soldier whose main error in life was a vanity that he cherished and displayed at the expense of others. Sergeant Major Hope, his guardian angel in this allegorical tale, tells him that he must atone for his overweening pride in his abilities as an airborne ranger. Because of it, says Hope, "You kind of put a whole bunch of other people down . . . you just didn't let them think that they were being all they could be, and if they were, it just wasn't anything to write home about."

The lieutenant colonel takes on his mission and leads his ghostly battalion, resolved to win the war of the training center and reach his heavenly goal, but he has a lot to learn. He suffers defeat after defeat, but he carefully and thoughtfully reviews his errors (helped by the relentless pressure of implacable observers who make him and his troops point out and acknowledge their many mistakes). Slowly A. T., who knew very little about the life of armor and mechanized units and their tactics and logistics, becomes a new man, a wise and canny leader. He and his unit grow with each new dose of adversity until he begins to turn the tables on his powerful enemy.

Then comes Hill 781, his last chance to show that he can accomplish his mission. In the culminating battle of this captivating story, he swallows his pride, the cause of his troubles, and in pulling together all that he has learned, he finds that success lies not simply in absorbing tactical lessons, but also in achieving the high level of understanding and respect for his subordinates and his troops. This recognition gives him the insight to allow them to participate fully in the planning, in the actions that follow from the plan, and in the individual initiatives that come from knowledge of the commander's concept. Confidence, empathy, and team spirit now build to a high pitch in the chain of command and in the troops, and the story ends in a crescendo of combat.

This fine book, action-packed and liberally illuminated with battle sketches both verbal and graphic, is deceptively easy to grasp; it is only after the reader closes the book at the finish that the full complexity comes home. In the end, A. Tack Always, the lieutenant colonel made new, has "earned the right to leave Purgatory."

And in the end Jim McDonough has earned every right to say that he has written a superb little classic of a book—one of the best on modern training that can be found today by anyone who is searching for some idea of the true lessons to be learned in the tactics of high-intensity, high-technology warfare—and the way to learn them.

John Galvin
General, USA (Ret.)

Preface

Shortly after commissioning in the United States Army I was sent to the Infantry Officers' Basic Course at Fort Benning, Georgia, in those days a twelve-week course designed to prepare the new lieutenant for his first assignment. There was little reading in the course other than the official manuals that explained basic doctrine, weapons employment, and administrative functions the young officer would most likely face in his initial post, most probably as a rifle platoon leader.

One book that was passed around, however, was an enjoyable little tactical primer entitled *The Defence of Duffer's Drift,* a fictional account of a British subaltern in the Boer War who sets out with his platoon to defend a lonely ridge line deep in enemy territory. Lieutenant Backsight Forethought, a pseudonym for the author, Ernest Swinton, is a novice at the business, but through a series of fortuitous dreams that reveal to him the error of his tactical ways, he awakens to the harsh realities of combat and sets in the defense properly by the time of his seventh dream.

The book was highly entertaining. More importantly, it imparted a number of tactical messages in a readable and unforgettable format, which stood me in good stead as I set out upon my own combat missions in Southeast Asia. Years later as a battalion commander I made it a habit to pass out copies of the book to all newly assigned lieutenants with the hope that it would help them to think hard about the tactical problems they might one day encounter.

It was during those years in command that I began to go through a number of training experiences at the U.S. Army's National Training Center, a rugged and realistic post in the Mojave Desert, not too far from Death Valley. It is here that our army has been practicing and refining tactical doctrine while rotating brigades and battalions through exercises that closely resemble the stress and strain of combat. Results of engagements are visually and electronically observed by the cadre, briefed back to the rotating units which simultaneously engage in a self-critique of their performance. The experience is revealing, usually humbling, sometimes shocking, and always educational. This training has done much to focus the army in the last quarter of the twentieth century and has helped prepare the American army to face the rigors of war.

It occurred to me after several exposures to the National Training Center and a little prodding from my friends that a book that passed on some of the lessons learned there might be a useful thing for the military profession, and of interest to the general reader. I have modeled this attempt to do that on the style of Ernest Swinton. In this case the lieutenant becomes a lieutenant colonel, and the rifle platoon a mechanized task force—a battalion-sized unit combining tanks, infantry fighting vehicles, infantrymen, mortars, scouts, engineers, air defense elements, their organic supporting elements of mechanics, cooks, truck drivers, supply personnel, administrators, medics, communicators, and the direct support of designated artillery, air cavalry (helicopters), and fixed-wing close air support.

Although the rank is higher and the unit more complex, the essential elements are not dissimilar. The leader must consider his opponent and his own resources and find the courage and the wisdom to overcome the former with the latter. Ultimately, he must recognize that his prime resource wears a human face and thereby apply the leadership that brings victory. In this may lie a basic truth of war through the ages.

CHAPTER 1

First Impressions

Lieutenant Colonel A. Tack Always found himself standing on the hot strip of desert sand that separated the endless straight track of the Santa Fe railroad from the dilapidated, broken blacktop road that accompanied it along its length as it disappeared in either direction over the horizon. A few dozen meters beyond the road lay U.S. Route 15, the major highway from Los Angeles to Las Vegas, over which traveled the eager souls hell-bent on throwing away the riches they had reaped from their industry in the lands astride the Pacific shore. His eyes were glassy, his head ringing, his battle dress uniform dusty and wrinkled, faded by long days and nights of unbroken use. For the life of him he could not remember how he came to be here, alone, unaccompanied by his soldiers, and without any means of transportation. Confused and befuddled, he walked over to the blacktop secondary road, trying to get a fix on his location. He had seen this place before, but that thought came to him only as if from a distant dream, unclear, hazy, and ominous. Where could he be? Why was he here?

Aside from the cars rushing past on the highway, there were no signs of life anywhere. Large power lines strung off into the distance, but nary a bird, jackrabbit, or snake broke his solitude. He was alone, utterly and completely. The heat was stifling and he turned to his canteen for relief, only to find himself choking down a stale, hot gulp of water.

I've got to collect my thoughts, figure out what the hell is going on, he thought, his mind virtually creaking at the effort it took.

As he walked toward a highway overpass several hundred meters away, he sensed a lightness to his body, oddly counterpoised by a heaviness to his soul. The thoughts just were not coming, and try as he might, he could not focus. A sign came into view as he closed on the overpass, taped to the columns supporting the roadway above, "HALT" emblazoned across it in big, thick black letters. Beneath, in finer print, was a series of instructions as to how a military convoy was to pass under, at what interval, at what speed, and so on, as if the poster of the sign was afraid that an unguided unit might sweep the columns out from under the highway, closing the artery bringing the sinners and their money to the Sodom and Gomorrah of the desert.

Well, that seems unfriendly enough, thought Lieutenant Colonel Always to himself. But the relative coolness of the shade beneath the overpass beckoned him on. For a moment he paused as his eyes adjusted to the darkened light. Then he saw a second sign hanging on one of the middle columns, this one less official looking than the first. Scrawled in an uneven hand with gaudy colors was the message, "Welcome to the Twilight Zone. Abandon all hope."

"Good morning, sir." A voice startled him from his reading, but with habit formed by many years he instinctively returned the salute as he tried to focus on the source of the words. "Are you Lieutenant Colonel Always?"

"Yes I am. Who are you, please?"

"Sir, I'm Command Sergeant Major Hope. I've been expecting you for some time now."

"You have?" Colonel Always was trying to gain his composure. Here was some hope that he might discover just what was going on. If a man, indeed a command sergeant major, was in this godforsaken place waiting for him, then there must be some logic as to how he came to be here.

"Yes, sir, ever since you died last night." The words hit Always like a thunderclap, and in an instant the memory came back to him—the long march through the swamps, followed by the steep climb into the high ground, along snaking ridge lines, rucksack knifing into his shoulders as he led his light forces for the umpteenth time on a field exercise designed to show their mobility, sustainability, and hitting power. The fatal step had come as a result of his own obstinacy, his decision to show his soldiers once and for all that there was absolutely nothing wrong with the army ration known as the Meal Ready to Eat, or MRE. How sick he had been of his men's derision of this space-age update of the old C ration, their snide referral to it as the Meal Ready to Excrete, and their utter conviction that the man was not yet born who could eat three of them in one day and live. And so it was with great fanfare that he ate one meal after another during the day, despite the warning signs that had been building throughout the afternoon—a reverberating wrench in his gut and a rumbling resonance in his bowels. He dug in, undeterred by the delectable delights of a barbecued beef. It was unclear if the final explosion was brought on by the dehydrated potato patty or the freeze-dried strawberries. All he could remember was his adjutant asking him if he would like some water to wash it down, his offhanded acceptance of that offer, a gulp, and a flash. That was the last thing he recalled before waking up here in the desert.

Fighting to retain his composure he asked, "Uh, look, Command Sergeant Major, I've had a hard few days and I would appreciate you refraining from any flip humor."

"I'm sorry, sir. I meant no disrespect, but I assure you

that what I say is true. You did in fact die last night and even now the accident investigation team is struggling with the problem of how to document the cause of death as the MRE, a completely unacceptable finding for the board. I can imagine how hard it is for you to accept, this being your first time dead and all, but I swear that it's true by the proof that you've left no tracks in the sand."

Slowly, Lieutenant Colonel Always turned his head back to glance from whence he had walked and to his dismay saw that he had in fact left no depression, despite the softness of the sand.

For the first time since he had found himself on the desert floor, a coldness swept over his body. So he was dead! The thought was heart-stopping, or would have been, he reflected.

"Well, if that's true, why are you here and where am I?" Always turned to Hope, afraid of what he might hear.

"You are at Manix railhead," said the sergeant major, indicating a lone cement ramp rising from the sand at the end of a rail siding. "This overpass we are standing beneath marks the entrance to what in life was known as the National Training Center. I am to be your guide in your sojourn here. We will be moving up the desert trail a couple of dozen miles. They're expecting you there."

It was all coming too fast for Always. He had steeled himself to accept the fact that he was dead, but what did the National Training Center have to do with that, and who was waiting for him twenty some miles deeper into the desert? He had been a professional officer for most of his adult life, and not a bad one at that. For the most part he had lived a decent and respectable life. Yet if these were the gates to heaven, and the man standing in front of him was the gatekeeper, it was not exactly what he had been led to expect.

"Sergeant Major, if you don't mind, I would like to ask a kind of personal question."

"Not at all, sir."

"Are you dead too?"

"Yes, sir, I am. Been dead quite some time as a matter of fact."

"And does that give you any insight into what this is all about?" Always was starting to regain some of his authoritative bearing.

"Well, Colonel, it does and it doesn't. I know this isn't all clear to you yet, so perhaps I should do a little explaining. The first thing I would like to make clear is that I asked to come down here, specifically to be your guide."

The word "down" gripped the officer in an icy vise, his breath escaped him, and for a moment he thought his knees would buckle. Could it be the worst had happened? What had he done to deserve it? Hadn't he always made the morning run with his troops? Never once did he tamper with a readiness report. And the annual general inspections—he had always pulled those off pretty well without undue harassment of the soldiers; well, at least without *extreme* undue harassment of the soldiers. And all the social events. Sure, he never liked them, but he had gone, behaved himself reasonably well, complimented his hostess on something or other in every case, and always tried really hard to make that one brilliant statement that would indelibly imprint itself on the minds of his superiors for later recall.

"Just what is it you mean by 'down here,' if I may ask?"

"Yes, sir, you certainly may. This is kind of a touchy subject for an enlisted man to be telling an officer, but the fact is, Colonel, you didn't quite make it into heaven."

As Always blanched at the words, the command sergeant major picked up his mood and quickly went on. "Now don't go jumping to any hasty conclusions. It's not as bad as you're thinking. If you didn't make it to heaven, you didn't quite end up in hell either." A sense of déjà vu hit the colonel as he remembered his last efficiency report. "The truth of the matter is that you've made it into Purgatory, which is what the National Training Center is used for. You see, sir, you didn't quite have an unblemished record in the army, so the System has arranged this little stopover for you until you can make it up. Just how long that takes is up to you."

Although the news was disconcerting, Always felt it was futile to resist it, afraid he might be left behind in the rush should he fight the logic of the words. The sergeant major was not being harsh, just straightforward. In that, Always found solace. There was something comforting about the noncommissioned officer, so respectful, so knowing, seemingly so in charge. It occurred to him that that was the way it had always been for him with sergeants major. It was a marvel how they could show deference to an officer, yet at the same time be so much on top of things.

Swallowing his pride, Always asked the burning question. "What did I do to deserve this? I mean Purgatory and all."

"Well, sir, I figured that would be one of the first things you might want to know, so I checked with the Chief before I came down here, and although many of these things are beyond me, I did get a feel for your particular situation. Again sir, meaning no disrespect, it had to do with believing your own propaganda, so to speak."

"Excuse me, Sergeant Major, did you say *propaganda?*" Always was clearly irritated at the pejorative term.

"Yes, sir, I did, but of course that's just my own word for it, and I can see it might not have been the best one. Maybe I can explain it like this. You know that army recruiting theme we adopted in the 1980s—'Be all that you can be!'—well, you started really believing that you had a corner on that market. Not that being infantry, and airborne, and a ranger weren't good things. In fact, that helped your ledger a great deal. But after a while you started thumping on that stuff a little too much, and, well, you kind of put a whole bunch of other people down while you were doing it, and when that happened, well, you just didn't let them think that they were being all they could be, and if they were, it just wasn't anything to write home about."

For all his faults, Lieutenant Colonel Always was an honest man, and even as the sergeant spoke he reflected on all of his disparaging comments about soft staff officers, "legs" (nonparatrooper qualified soldiers), support branch personnel ("remfs," "wimps," "pukes," et cetera). It was true. How much he had coveted his senior parachutist wings! How heroic it had been to posture about his ability to walk unlimited distances, suffer sleepless nights in the cold and wet, thump his chest and bellow the guttural sounds so endearing to all real infantrymen and so offensive to those who wished they were. But he had not thought there was anything wrong with that. After all, he had been taught the very same things when he was a young officer—unless, and the thought was sobering—unless those who had taught him had also ended up in a mess like this. Maybe he had embellished some of those war stories a bit too much, but to consider that to be anything worse than minor exaggeration . . . , well, that seemed a little hard. He hadn't meant to hurt anybody's feelings, even if they were miserable pantywaists. Always' head began to hurt from all this thinking.

"Sergeant Major," Always seized the initiative. "It seems to me that if this is Purgatory, and mind you I'm not convinced yet of anything I've heard, then I seem to recall that it's only a kind of transitory post, sort of a temporary duty station, until I can complete my business and get on to my permanent assignment."

"Right, sir."

"Well then, just what are my terms of duty here?"

"Do you mean how long are you here for and what do you have to do?"

"Precisely."

"As I said, sir, I know a few things about what's going on but not all of it. How long you're here for is up to you. The purpose, as the Chief put it, is 'to teach you the error of your ways.' "

"You mean I'm to be punished for my, uh . . . well, my 'arrogance'?"

"No, sir, not really punished. That's not the way the Chief works. It's more like He wants you to appreciate what some of those other guys, those guys you made a habit of belittling, do. He feels that you never will really have a place amongst them in heaven unless you first learn what an important contribution they make."

"And can you tell me, Sergeant Major, just how do I gain that appreciation?"

"Colonel, that remains to be seen. I've a general idea of the plan, but I really don't know all the details. To be quite honest, sir, I think I've done my job of bringing you in on the problem, and probably the best thing we can do now is to proceed further on into Purgatory, the National Training Center that is, and you can see for yourself what's in front of you."

For a second Always thought of overriding the sergeant major's suggestion, but there seemed to be an essential wisdom in the thought. For the moment he had absorbed about all that he could. Besides, he was tired of standing and, having always been an aggressive individual, he was eager to plunge ahead and see for himself just what was in store for him.

"Very well, Sergeant Major. Do you have a means of getting me there?"

"Certainly, sir. Even in Purgatory a lieutenant colonel is still a lieutenant colonel. I've brought a jeep. If you'll follow me I'll take us on in."

As the two men left the shade of the overpass, the hot desert air seemed to thicken. A dust devil ominously swirled up before them, and a fiery blast of heat and sand burned their faces.

Always hesitated only for a moment, then squared his shoulders and walked over to the jeep. Protesting at every bump, the vehicle slowly made its way across the sand-covered trail that wove through the mountain ranges ringing the vast, bleak expanses of Purgatory. He stole one last glance at the deserted Manix railhead as a Las Vegas–bound mortal sped his way across the overpass, rushing to sample the vices that lay before him.

The drive through the high desert was hot and uncomfortable, but it gave the two men a chance to talk. Lieutenant Colonel Always learned that, in recognition of his stature in life, he was to be given command of a battalion-level task force. Although it would be fundamentally a mechanized infantry battalion, he would have attached to it two tank companies to give him an armor punch, and would detach two of his own infantry companies so that they could be sent elsewhere to round out an armor battalion as a similarly tailored task force. Moreover, he would be given the most modern of equipment to work with, the M1 Abrams tank and the M2 Bradley infantry fighting vehicle.

That concerned Always a bit since he had made a career out of avoiding what was known as "heavy" forces. Showing a studied disdain for any soldiers who depended on machines to transport themselves, he had thus avoided the headaches that come with meshing men and machines in the business of soldiering. He once had been the executive officer of an airborne company, and the organizational procedures it had taken to keep the two jeeps operating had been maddening. With that experience under his belt, he had become convinced that there was no way he wanted to deal with the more than 200 tracked and wheeled vehicles found in a mechanized battalion. He would now be commanding two mechanized infantry companies, two armor companies, a mechanized antitank company, a reconnaissance platoon, and a mechanized heavy mortar platoon, as well as a maintenance element of more than 100 mechanics and heavy equipment operators, all supported by yet another platoon with a myriad of supply and fuel trucks. To make matters worse, all kinds of attachments would further stretch his span of control—engineers, artillery coordination teams, air defense soldiers, and air force liaison—each of them with their own particular items of equipment that needed to be supplied, repaired, fueled, and operated. If this wasn't hell, then it certainly was a nightmare.

"Command Sergeant Major, what kind of soldiers will I have?"

"Sir, let me assure you that you will have the normal type of soldiers you are used to, which means very good ones." The American army had always been blessed with capable soldiers, making a mockery of the claim—that so many of her foes delighted in rendering—that they were products of a soft and undisciplined society and could only crumble under the trials and sufferings of combat. "But like in any unit, sir, they will only be as good as you can let them be. Your staff knows the essentials of its job; your men are physically fit, dedicated, and technically competent; and none of them is reluctant to work hard. I assure you that the raw material is there, and I might add that the equipment, all of which postdate my own time on earth, is pretty good. Like always, though, how you put them all together will be the critical determinant in the overall fitness of the unit."

"Sergeant Major, what is their status? I mean, are they alive, or are they like me?"

"Colonel, they are just like you, here for a while until they can prove themselves and move on to a higher reward. As soldiers they haven't exactly been saints during their time on earth, but then again they have been a pretty decent bunch of guys. The Chief keeps a high set of standards, and although He's got a real soft spot for soldiers, He does have to let them work off their little faults. That's where you come in. You can help them, and they will help you. When you both get it right, you can expect to move on."

"You mean that how long I, rather we, stay here in Purgatory is dependent on what we do?"

"Yes, sir, exactly. The plan is really quite simple. You are to take command of the task force, commit it to battle against a well-trained enemy, and when you and your soldiers have defeated them you can turn in your equipment and move on."

Always turned the thought over in his mind. Things were not so bad after all. He was a professional soldier, and he had often been put to the test, both in actual combat and in years of peacetime training. Surely he could figure it all out pretty quickly and be on his way. He had done it before, and he would do it again.

As if anticipating his thoughts, the sergeant added, "There are a couple of hitches though, and it could be a little tougher than you think. First of all, the enemy plays by a different set of rules. You see, they've been here a while, the little devils. This is their turf, so to speak, and they know every nook and cranny out there. Furthermore, they get a break on resupply, rest, recovery, and reconstitution. They also don't get the visibility that you do."

"What do you mean by 'visibility'?" Always asked.

"Well, sir, since they aren't working toward the immediate goal of getting out of this place and on to better things, there's really no need to scrutinize their way of doing things beyond whether or not they've roughed you up in a fight. So none of the evaluators spend any time criticizing their techniques, double guessing their methods of operations, or otherwise adding torment to their condition."

"*Evaluators?* You mean I get evaluators? I thought that was what you were, Command Sergeant Major." Always was perplexed.

"Oh no, sir, I'm your task force sergeant major, like I said before, the only volunteer down here, and your guide through this ordeal. It's not my place to evaluate you. It never was in life, and I'm certainly too old to start doing that sort of thing now. But I can give you advice every now and then, when you care to hear it. That's why I volunteered for this job. Heaven's a nice place and all, but they never have any crises up there ever since Luther and the fallen angels got kicked out, and us sergeants major, well we kind of thrive on crises. So here I am."

"And the evaluators?"

"Well, perhaps I shouldn't call them evaluators. They are officially known as 'observers.' They kind of watch you and help to point out the error of your ways. You'll meet them soon enough. Nasty bunch, if I do say so myself. Must have been particularly rotten in life. They'll show up en masse as soon as you get your operations orders, follow you everywhere you go, say disparaging things to you, talk badly about you over their radios, and render a report as to how you did. The only saving grace is that they're being evaluated too, and if you don't show any improvement, well then it gets sticky for them. Some have even been bounced out of here to a lower level, like recruiting duty. But one thing is for sure, their time down here is much longer than yours, so don't expect much sympathy from them. As to their 'observations,' well, think of them as hometown judges in an away fight. If you don't knock out your opponent, don't expect anything better than a draw, and that only if you creamed the other guy."

Always began to feel a little ill.

"One other thing while I'm telling you all the bad news. They have a superb electronic setup down here. Everything you say over the radio will be recorded so you can't deny you said it later, and everything you do will be filmed so that your most ridiculous moments can be played back for all to see. At any time you can expect everybody and his brother to be eavesdropping on you, offering their views as to how incompetent you are, spreading disparaging rumors, and unequivocally stating they could do it better."

"Sergeant Major, that doesn't even sound decent. It sounds like the only escape I'll get from all this misery is when I'm sleeping."

"That's the hell of it, sir. You won't be sleeping down here. Oh, you will be told to get some sleep, in fact you will be seriously chastised for not developing a 'sleep plan.' But if you should ever get some sleep, the observers will devise a scheme to wake you up so they can point out how things fell apart while you were sleeping. After that they will point out how things fell apart because you didn't get any sleep."

Always groaned and the conversation drifted off as they made the last leg of the journey up from Langford Lake (dry as a bone) to the outer ring of Purgatory known as the Dust Bowl.

The scene was utter bedlam. Thousands of troops were hurrying to and fro, jumbled up amidst countless vehicles, their diesel engines making horrendous noises, filling the already hot, stifling air with ever more hot and noxious fumes. Dust and sand blew every which way, pelting the soldiers' faces, covering them with a gray mask from which protruded sun-blistered noses and chapped and peeling lips. Red-rimmed eyes revealed an intensity of purpose forlornly trapped in hopeless frustration.

"What's going on here, Sergeant Major?"

"This is the equipment draw, sir. The battalions reporting in are being issued their vehicles. For several days the troops are indoctrinated to the hellishness of this place. They bivouac here in this flat, open expanse where there's no shelter from sun, wind, sand, or dust and spend their waking moments at the mercy of the ghouls who issue the equipment. The latter are a particularly nasty lot who cajole the men into drawing badly worn equipment, telling them that it's really in good condition. The catch, though, is that they make them sign for it with terms that they can never leave this place unless it's returned in the good order in which it is drawn. Of course, the condition when drawn is overstated, so that the possibility of ever getting it to the stated condition is remote."

"If that's so, why do the men ever sign for it? Surely they are experienced enough to recognize worn-out equipment when they see it."

"Yes, sir, they know it. But they really don't have any choice in the matter. They've got to draw the equipment so that the battalion can go into battle, so it can be tested, so it can defeat the enemy, so they can get out of Purgatory and on to heaven. If they don't accept the equipment, then none of the rest of it can happen, so they're under tremendous pressure. But when they go to turn it in, they are obligated to restore it to near-impossible condition. Until they do that, they can't leave here."

"Why that's fiendish, Sergeant Major. I guess at least some good comes of that—when the next battalion of lost souls reports in, they have good equipment to work with."

"Oh no, sir. Part of the contract the ghouls have with the managers here is to sabotage the equipment between turn-in and issue. That way the next battalion gets the same rotten stuff the preceding one did. That has a twofold impact. First of all, it adds to the reputational damage of the latest battalion through. Second, it almost ensures breakdown of key equipment at critical points in the battle, giving the local enemy the edge."

Always made a mental note to go easy on the mechanics. If they were the good soldiers that Hope said they were, they would be nearly killing themselves to keep the vehicles in running order, all the time bearing the brunt of the blame for operational failures. Despite his career with foot mobile forces, Always had been around the army enough to come across many of these poor overworked grease monkeys, as they were called, sodden with diesel fuel, lying under a multiton steel machine in a puddle of grime and filth, turning a wrench with sore and scraped hands, while officers admonished them to hurry up and get the cursed vehicle back into operation and to get busy on the next five broken machines that were waiting for repair somewhere down the road. Theirs was a particularly hallowed commitment, and to punish them further in a place like this just didn't seem right. He resolved to support them as best he could. After all, given his limited experience with heavy equipment, he would be dependent upon their knowledge and dedication to pull him through this trial.

The jeep pulled up to a cluster of tracked vehicles with canvas extensions protruding from their rears, all joined with a series of snaps and supported with a few poles so that a temporary shelter was formed at their confluence. The side flaps were rolled up to let some of the desert air flow through the shaded area beneath, and inside Always could see an intense group of officers bent over maps and talking on radios.

"This is your task force headquarters, sir. The executive officer is out with the supply and maintenance people right now, but the operations officer should be here. They've been expecting you. I'll be leaving you here for a bit, sir. I've got a few duties to attend to while the staff briefs you. I'll be back after the briefing to take your instructions. Good morning, sir." The colonel and the sergeant exchanged salutes.

As Hope pulled away it occurred to Always how much he liked the man. As bad as things might be, with a command sergeant major like that on his side he could not go too far astray. He would have to be careful to ask for and to adhere to his advice, Always reminded himself.

"Good morning, sir. Welcome to the Dust Bowl. I'm Major Rogers." Always turned away from the departing jeep to face his operations officer, a solid, open-faced man of medium height.

"Good morning, Major. Pleased to meet you. I'm LTC A. Tack Always, your task force commander."

"Yes, sir, we've been expecting you," Rogers answered cheerily. And on that note the two men proceeded into the tactical operations center (TOC), where Lieutenant Colonel Always was introduced to his staff and given a series of briefings on the state of the command and the missions before them.

Purgatory had been well chosen. The land is hard; the climate is brutal. The high desert rises to an elevation of three to four thousand feet on the valley floors, and many times higher on the barren mountains reaching toward the sky. The ground could change abruptly from the dry lake beds of cracked and parched earth covered in areas with a fine, deep dust that engulfs all trespassers in a choking cloud of filth to hard lava rock that could just as readily rip off a man's boot at the soles as strip the rubber pad from a tank tread. Everywhere the ground travels in crazy angles, sloping up toward the higher ground, then abruptly away toward the dry streambeds known as wadis. Hulking mountains turn into sheer cliffs, offering elusive passes to other severe terrain features. Estimations of distances are maddening in the clear desert air, things seeming much closer than they actually were. It would be easy for a gunner to fire at a target hopelessly beyond the range of his weapon, believing all the time that it was well within the envelope.

At night the ground is engulfed by an eerie darkness, made more desolate by the vast emptiness of the endless desert. Without a moon to light the way a man could easily drive off the desert floor into one of many deep and treacherous wadis that crazily zig and zag across the valleys, unseen from any distance whatsoever, even in the brightest sunlight. The wadis offer a perplexing tactical problem, often suggesting a convenient avenue of approach to an armored column only to narrow to an impasse from which the only escape comes by backing up, a treacherous undertaking for a column of vehicles. The desert floor itself is anything but flat, a series of waffle-like rises and depressions that knock the senses out of a man and the mechanisms out of a machine.

Accentuating the rigors of the terrain is the unrelenting weather, either too cold or too hot, too stifling or too windy, sparing nothing that painstakingly makes its way across the map of the desert. The sun rises suddenly, mounting in oppressive heat throughout the long summer so that by nine in the morning a man begins to yearn longingly for the end of the day. There is no shade, and as the sun rises higher in the sky its rays beat down ever more directly until by noon any metal surface—a weapon, a tank, a jeep hood, a helmet—is burning to the touch. By midafternoon men are going mad in the heat. Heads ache and eyes are skewered by the glaring rays of sunlight reflecting off the sand, a condition that worsens as the sun sinks lower in the sky and blinds any living thing trying to make its way in the direction of the setting sun.

In the winter the sun rises just as surely, but only to mock the inhabitants of the desert with its refusal to warm, burning only the eyes of the disappointed beholder. Then the nights become almost unbearable as the meager heat of the day escapes into the unclouded atmosphere, leaving the unsheltered below chilled and shivering in the long night. Whatever heat has been gained serves only to accentuate the coldness that follows.

And the wind—winter or summer—is merciless, drying, cracking, sandpaper-like with its airborne residue of the desert scraping away live skin from man and animal alike. When the blowing starts, it is relentless—cascading down the broad valleys from the high mountain ranges, building in fury until nothing can stand before it. Tents are ripped asunder, equipment is sheared from the decks of vehicles, men grope for cover as the needlelike sand fills their eyes, their noses, their mouths, their lungs. Speeds pick up from twenty miles per hour to near one hundred as the storm builds hour by hour, never slackening, seeming to be an endless trek of mounting violence and destruction. Humans choke and vomit from the shovelfuls of dirt that smash into their faces, now bleeding from the unceasing sandblasting. In the heat of summer the dehydration effect becomes life threatening. In the cold of winter the windchill factor plummets to extreme subzero readings, making frostbite an ever-present threat.

Then suddenly the wind dies and the eerie quiet of the desert returns, the landscape somewhat altered by the shifting sand but still identifiable by the stark terrain features of rock piles turned mountains as if by the idle play of some giant's hand. And somewhere in that fearsome playground lurks an enemy, an enemy familiar with every valley, pass, crest, wadi, and crevice, a familiarity burned into his brain with constant and repeated exposure; an enemy wily in using the terrain to every possible advantage, who knows the line of sight of every weapon's emplacement, who can register the fires of his artillery and mortars by memory; an enemy who has seen a score of times the blunders of his foes led astray by the deceptive terrain as they followed wadis into disastrous death traps or tried to scramble up seemingly gently sloping paths that suddenly turn into unclimbable cliffs; an enemy who knows every hidden crack in every cliff face, who can dig in his vehicles and men so expertly that they cannot be seen within spitting distance, but who can see out to the limits of their ranges and beyond.

This was what Lieutenant Colonel Always learned from his staff in his initial spate of briefings, and what he would soon see for himself. The briefings were good, professional, succinct but to the point. They ran the gamut from enemy and terrain to maintenance and medical status of his own equipment and troops.

One by one, his staff officers stood before him and outlined the problems that lay before him and the resources he had to overcome them. They spared him nothing and answered his questions straightforwardly.

"When will we get our first mission?" Always asked Rogers.

"Sir, it will be received tonight when the observers descend on us for the first time."

"From what I've heard, it figures they would wait until dark," Always quipped. "Arrange a meeting here for me with my commanders at dusk, and have a helicopter standing by at first light so that I can make a reconnaissance in the morning. I need to see the ground for myself. I'll be leaving now to find the executive officer. Continue with your contingency planning. Oh, by the way, I understand that I will be leading the task force from my own Bradley, my infantry fighting vehicle, or IFV as they say. Get my crew here with the vehicle so I can meet the men and learn a little about the machine before dawn. Any questions of me before I move on?"

"No, sir. I'll line up all of that."

"Very good. I'll see you later."

Lieutenant Colonel Always greeted his jeep driver and sped off to find his executive officer in the jumble of activity over by the vehicle park. His dread of motor pools was soothed only by the good-naturedness of his driver, Specialist Sharp, a tall, alert-looking young man, who despite his attempt at an efficient but disinterested manner could not conceal his great interest in sizing up the new commander.

"Well, Specialist Sharp, what's the word on this battalion I've inherited?" Always was probing, knowing full well that in the next few minutes he would learn a great deal about his unit and his men. So far he had been impressed with the battalion; and the clean shavenness, brisk manner, correct salute, and open honesty of Sharp had already impressed him further. It was amazing how quickly a military unit could reveal itself in its men and its equipment, and very few of them at that.

"We're a pretty straight unit, sir. We make mistakes every now and then, but with a steady hand from the commander we'll do just fine out here." Always was now even more impressed by this stalwart soldier. A quick look around told him that his words might be on the mark. The soldiers within sight looked like soldiers should—uniforms worn properly, noncommissioned officers in sight everywhere supervising their men, officers and enlisted men exchanging salutes with obvious mutual respect, equipment seemingly cared for (proper bumper markings, vehicle canvas stitched where needed and tied down firmly, weapons clean, et cetera). Even the chatter over the radio was correctly done, crisp and to the point.

"Over there please, Specialist Sharp." The colonel indicated a commanding figure standing amidst a line of tanks being inspected by the ghouls issuing the equipment.

"Good afternoon, Major Walters." Always surprised his executive officer, but only for a second.

"Good afternoon, sir. Major Rogers just called on the radio and said you would be coming over. We've got things under control, sir. Here is the equipment readiness report." He handed Always a copy of the list of inoperative vehicles and radios, keeping the original for himself. "I can brief you on when we can expect them to come up, if you would like me to begin there."

Always exited the jeep and found himself looking up at the tall major. He was glad to see his XO so abreast of the maintenance status of the unit, but he asked him for a larger view of the battalion, about morale, personalities of the staff and commanders, the key noncommissioned officers, the state of readiness, discipline, military bearing, and courtesy. He noted approvingly

Walters' care to defame no one in his initial discussions of them. Even more admirable was Walters' reluctance to pass on any judgment on the commanders. By so refraining he showed his keen awareness that relationships among commanders are special, and that a staff officer has no business trying to formulate what that relationship should be. As is always the case, by his choice of words Walters was saying much more about himself than about the unit or anyone in it.

Always marveled at what he was discovering. Here was a unit as well disciplined, as well manned, and apparently as well led throughout its subordinate elements as any commander could hope to expect. It occurred to him that he would have no excuses for whatever ill might befall the task force in the operations to come. If he could not put this task force together into a coherent whole, it would be his own failing.

Throughout the day, as Always moved around looking into every nook and cranny of his task force, talking to soldiers, meeting officers and sergeants, seeing and being seen, the thought was reinforced. His sense of identification with the battalion deepened, and his preoccupation with his own predicament was supplanted by a preoccupation with the unit. He was careful to reinforce the strong qualities of discipline and self-respect he found everywhere. He found himself straining to look the part of a strong commander, to sound the right note of firmness and encouragement in his conversations, to flatter when it was due, and to correct, but not harshly, when it was warranted. He felt the men respond to him, the almost 1,000 of them who constituted his task force with all of its attachments, as if they picked up his personality, and he theirs. By evening his desire to not let them down had surpassed his desire to get himself out of Purgatory. If he achieved the first, the second would follow, but it was the first consideration that stressed itself to him. Concern for his unit and its people had overtaken his concern for himself.

Somehow, the burden of the responsibility eased the disquiet Always had known since awakening that morning at the gates of Purgatory.

As Always arrived back at his TOC, Major Rogers hurried out to meet him. "Sir, we've just received our warning order from Brigade. We'll be making a night road march tomorrow night up to a tactical assembly area south and west of Hill 931, to be followed by a dawn attack onto Objective BLUE, north of Hill 826."

Always looked at the map. Plenty of time, he thought to himself, and the terrain doesn't look that rough, at least not from the map. "Do we know what's up there?"

"No, sir," Rogers answered. "The S-2 has put in an intelligence request to higher headquarters. We're hoping to get a quick answer to that question."

"Very good. Work me up a road march order. The attack doesn't look that tough. After I meet with the commanders tonight have the XO get the staff together and work me up some options. We'll mull things over, make a decision, and get the word out to the subordinate elements in the morning. In the meantime, keep that helicopter on order for me so I can make an early reconnaissance."

"Very good, sir. By the way, our observers are due momentarily."

A cold chill went down Always' spine. During the afternoon he had become wrapped up in the myriad of details of commanding a battalion readying itself for action. Now he was reminded that this was to be no ordinary operation. A commander enjoys being king to his own soldiers. He didn't want any godforsaken souls coming in to throw their weight around in his unit.

But wanting and getting are two different things, and in an instant the roar of scores of jeep engines and a wave of dust engulfed the placid scene of operations officer and commander having a civil discussion. Like so many jackals, the dreaded observers descended upon the headquarters, each one seeking his counterpart, with a sneer upon his lips and an air of contemptuous disdain for the hapless victims. A harder bitten lot would be difficult to imagine—faces seared by the desert sun, eyes glaring with sadistic eagerness, hands calloused and chapped from the writing of so many long and derogatory reports.

In their midst strode the most savage looking of the lot, a bull-necked demon emitting unmitigated callousness.

"Are you Lieutenant Colonel Always?" he bellowed.

"Yes," answered Always, trying to deepen his voice and sound unintimidated.

"I'm Lieutenant Colonel Drivon. I'm here to help you." The dreaded words passed through his thick lips with a menacing, guttural snarl.

"I'm glad you're here." Always was trying to hold his ground.

In such manner the two of them sparred for several minutes, but it was clear that Drivon had the upper hand. Already his assistants were cornering their victims, admonishing them to attempt no subterfuge, to confess their sins openly, to display their mistakes unashamedly for all to see, and to appreciate gratefully all the wonderful advice they were about to receive from their benefactors (read "observers").

Always made a mental note to settle down his people later from this disquieting experience, to point out that there was no good to come of resistance to the presence of the observers.

"Let me see what you've done with the warning order you just got." Drivon snapped Always from his thoughts, and he passed on what little he had to show.

The silence was deafening. Immediately Always wished he had pulled out the approved staff manual and gone down the checklist step by step when his operations officer had first approached him with the news of the order. But there had been so much to do, and it had seemed premature for him to go too far into the planning until he had formally met his commanders, familiarized himself with his battalion, and learned his equipment.

Now he found himself trapped by Drivon, who proceeded to consume the next hour and a half explaining the ground rules of their newly founded relationship. Each staff observer then added his own views as to how his part of the operation was to be conducted, and the time stretched out well after dark. Then, as suddenly as they appeared, the observers evaporated into the night, leaving Always to try to gather up his staff, call in his waiting commanders, and recapture control of his own headquarters.

At that moment he spied Command Sergeant Major Hope. "How are you doing, sir?"

Always was glad to see him, his polite, smiling, knowledgeable face relieving in an instant the turmoil Always had been suffering through for the last few hours.

"Sergeant Major, don't you have a counterpart observer?"

"Oh no, sir. I'm above all that, if you remember. Besides, no one has ever dared to dictate what a command sergeant major is supposed to do, so these guys down here would be out of line even trying to tell me. No sir, I neither need nor want their advice."

"God, Sergeant Major, I envy you!"

"Yes, sir. I know what you're going through. Just take it in stride. Remember, no one said this was going to be easy."

By now the officers had assembled, and the two men broke off their quiet conversation and moved to the front of the group. In unison the men snapped to attention, commanders up front in the lighted TOC, the staff and specialty platoon leaders to the rear.

Major Walters now took charge of the meeting, and one by one brought up the staff officers to brief the commanders. Most of the discussions had to do with administrative details, since Always had deferred any discussion of the order until later.

By his mannerisms, by his questions, by his kindness to the briefers when they were forthright and by his harshness when they were defensive, Always began to project his personality over the assembled group. A glance at Walters told him that his XO would be giving stern instructions to those briefers who had failed in the need to be thorough but concise. Again, Always was reassured at the professionalism of the group he had inherited.

The commanders were a good-looking lot. A and B Company commanders, Captain Archer and Captain Baker, were aggressive, intelligent young men. They exuded confidence and strength; they were at ease in talking of their men and their equipment and in stating their needs, but they also exhibited an understanding of the needs of the larger unit. Both their Bradley companies, Always learned, consisted of thirteen Bradleys each, but only sixty dismounted infantry soldiers when at 100 percent strength. Three men would have to remain with each vehicle in order to keep it moving and shooting when the infantrymen dismounted. Always was aghast. He had an infantry battalion with very few infantrymen.

Captain Carter of C Company was a short, solid tanker. His mannerisms indicated a man who was on top of every issue in his unit, almost artificially so, Always thought to himself. D Company's Captain Dilger was a lanky Southerner, slower of speech than the others and seemingly unassuming, but there was a steeliness in his eyes that was reassuring. His fourteen tanks, Always imagined, would be a heavy punch when the time came. Added to the fourteen others under Carter, that punch should be unstoppable.

Captain Jim Evans had the antitank element, the smallest of the five combat companies, but as he spoke of his unit in his New York twang, it was clear that he was confident it was a formidable force. He had only two platoons of missile carriers of four each, his third platoon having been detached to the armor battalion working on Always' flank.

The sixth commander, Captain Coving, was the most experienced of the lot. He had to command more than 300 men in Headquarters Company, mostly supporters—mechanics, staff, cooks, medics, and administrators—but he also had the mortar platoon of six tubes mounted in tracked vehicles and the scout platoon mounted in six Bradleys. Coving would have to possess great versatility to keep all of these elements pulling together in support of the upcoming missions, but he looked like he had the character and the skills to do it. Intelligent eyes accentuated a rugged appearance. He listened well and had the answers to a myriad of detailed questions that came up during the briefing.

If anyone ever doubts the strength of America, Always thought to himself, he should look into the faces of the young captains that come to command in our combat forces. It seems unbelievable that a nation so steeped in hedonistic values could produce such hard-working, self-sacrificing men, so physically and mentally tough. The demands on them were impossible. Responsible for everything their men did or failed to do, liable for millions of dollars worth of equipment, vulnerable at every moment to mishap, misfortune, or misdirected orders, they nonetheless approached their duties with a total commitment and dedication, working hours that would fell an ox under conditions of physical discomfort that would crack the resistance of lesser men. Yet there was never a lack of willing candidates to step forward and pick up the yoke—enthusiastic and zealous young men, aware of the deep responsibilities they shouldered for their nation. Always had seen endless thousands of them in his lifetime, and it never failed to inspire in him a tremendous pride in his country and his profession. Among the many great resources America can claim, surely its capacity to produce such valiant men ranks at the top of the list.

As Always took the floor to state his philosophy of command, to stress those values he would hold central to the men under him, he tried hard to establish strong eye contact with each of his commanders. Gathered inside the small canvas-enclosed operations center were the key men of the battalion. Clearly the most crucial to the health of the command were the company commanders. He had to know them intimately, and they had to know him. Each had to be attuned to the others' way of thinking. And so it was to them that he directed his comments.

"It is clear to me that this is a fine battalion. Your professionalism, your energy, and your strong leadership have been apparent throughout the unit. I can think of no outfit I would rather find myself with on the eve of battle. I lack experience in fighting a heavy force, but I know that with you to help me I will learn much quickly. I assure you I will commit myself to that end. In the meantime I know the bedrock of any unit is discipline, and that it is present in this battalion in ample quantity. We shall build upon that strength, as it will become more and more crucial as we face the tasks awaiting us.

"I have yet to meet a man in this outfit whose heart is not in the right place. We are all trying to do our duty as best we know how. That is all I ask of you. It can be expected that we will make mistakes; good men trying to do the right thing often do slip now and then. Don't worry about that. I am sure I will make mistakes too. I trust you will be tolerant of me, help me to correct them, and to avoid repeating them a second time. I will work toward the same end with you. In the meantime, our purposes are best served with mutual trust and honesty.

"Our primary mission here is to defeat the enemy. He is reputed to be a formidable foe. So are we. The men I met today did not lack for courage or commitment. With them we will be able to rise to every challenge before us, no matter how tough it is. We owe it to those men to give them the very best leadership we can. I am sure they will not let us down. We must not let them down. Tomorrow you will be issued your first operations order. I look forward to going into battle with you."

Always hoped that he had shown the right sense of balance in his few words. He knew that any insincerity on his part would have been perceived instantly, so he had spoken what he felt. Even at this late hour the impressions he made on this group would permeate the battalion, building on perceptions of him garnered earlier in the day. The confidence with which this battalion went into combat would grow from those perceptions.

It had been a long day. Always gave a few quiet words of instruction to his executive officer and his S-3, then went out into the dark to meet his Bradley crew and learn what he could of this unfamiliar piece of equipment before he would have to take it into battle.

CHAPTER 2

Dawn Attack

Dawn came early and abruptly in the desert. By four in the morning the sky had lightened, revealing a scene of continuing activity as the battalion finalized its preparations. Well before five the colonel had departed in his helicopter, flying over the route they would take that evening to the assembly area for the following morning's attack. Before he had left he gave instructions to his staff emphasizing the dispositions of forces for the attack and the dispatching of the scout platoon to make a reconnaissance of the route to the battle area. The order would be given at noon when the commanders came in to the tactical operations center. Things seemed to be going well, only the ever-watching observers lending a disconcerting air to events. By this time they had saturated the battalion, from headquarters down to each platoon.

From the air the desert floor looked exceedingly flat. The assembly area itself was well placed behind a convenient piece of high ground, shielding it from the observation of the enemy and affording a number of routes into the objective area. Always wanted to fly closer to the objective, but knew that to do so entailed great risk from antiaircraft fire. So he contented himself by orbiting ever higher to get a good standoff look into the area. On the flight back to the headquarters he could see the scouts moving down the route, fanning out to check the points that threatened the safe passage of the battalion's units. The battalion commander considered the risk of one of the elements in the column making a wrong turn in the dark, and so made a note to have the scouts man difficult points on the route in order to guide everyone on their way. He could sense the great risk that the nighttime movement of more than 200 vehicles and 1,000 men entailed. He was concentrating hard on the control measures he could adopt to keep everything on track.

Somehow time seemed to be getting away from him. As best he could tell he had not wasted any of it, yet as he discovered more and more things he had to do, the less time remained for him to do them. He wanted more time to study the map, to consult the doctrinal manual on road marching and attacking with a mechanized task force (the manual had chapters for every conceivable type of mission a task force could get), to familiarize himself with the equipment, to check on the progress in the battalion, and to prepare personally his equipment (maps, radios, gear, vehicle, et cetera), but he was continually distracted by unexpected events. His staff, diligently trying to get out a proper written order, was demanding an early briefing to him so that it could produce the final copies complete with all annexes by the noontime order. The staff officers kept coming to him for decisions: How many medics did he want to assign per company? What time would he like to feed the battalion? Should it be hot, prepared rations or MREs? (This gave Always an instant stomachache as he remembered what had gotten him here in the first place.) Where was his command post going to be? Whom did he want with him in it? What would the uniform be? What time did he want air support? (They had to plan it twenty-four hours in advance.) What would the order of march be? What missions would they give to the engineers? When did he want his tent struck? Could he meet with the Brigade S-3 and S-2 who had come by for a staff visit? Did he know about the difficulty they were having with tank transmissions in C Company? What mix of antitank missiles did he want in A and B Companies? Did he want to split his mortar platoon into two sections or keep them intact? How many radios did he want in his Bradley? What channels did he want them preset on, and on and on and on. It was not that his subordinates were incompetent—they only wanted to be sure to incorporate the wishes of their new commander. So valid questions were mixed in with trivial ones, and the colonel developed a searing headache trying to deal with them all.

By the time the orders briefing rolled around he had yet to break free of the entangling morass and reflect on just what it was he wanted to do. The result was an order that, although adhering to the staff manual format, was far from his own. Nor was it the expression of any clearheadedness on the part of the staff—each of them had been trying so hard to fathom the intent of the commander that any continuity of thought had been lost. The disaster was clear to the commanders present, when in order to understand their own role within the intentions of the battalion they were compelled to ask a plethora of questions that should have been resolved by the briefing. It was a tribute to Always that he did not lose his composure; he was experienced enough to know that this would only exacerbate an already bad state of affairs. He tried his best to clarify where it was needed, and this took up so much time that he feared his subordinates would not be able to do their own planning before darkness overtook them and they were caught up in the execution of the operation.

As it turned out he stressed the more immediate half of the operation, the road march, and he tried to overcontrol the movement, which revealed his inexperience with mounted operations. A short discussion was held in regard to the deliberate attack that was to commence at dawn, suggesting that planning to be

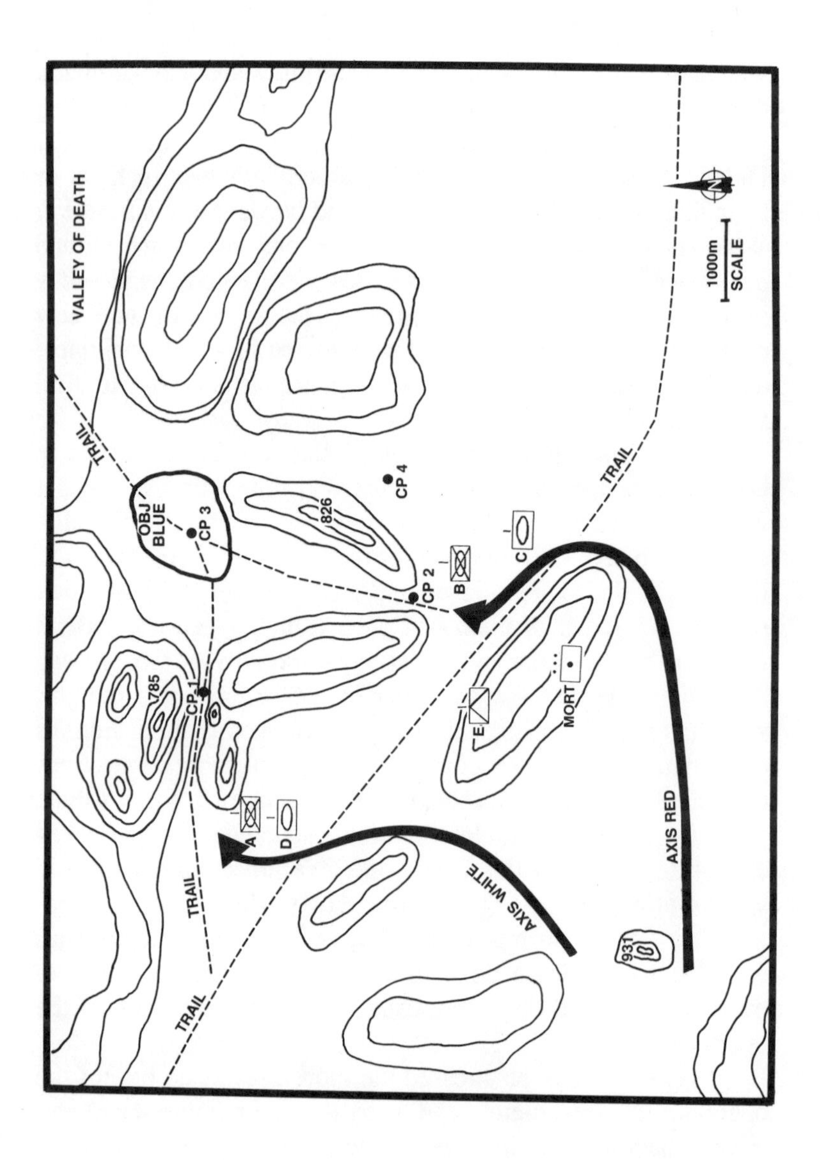
VALLEY OF DEATH
TRAIL
OBJ BLUE
CP 3
826
CP 4
CP 2
B
C
TRAIL
1000m
SCALE
785
CP 1
E
MORT
A
D
TRAIL
AXIS WHITE
AXIS RED
931
TRAIL

Map 1. Attack on Objective BLUE

B Company	13 Bradleys (M2); + 60 infantrymen*
C Company	14 tanks (M1)
A Company	13 Bradleys (M2); + 60 infantrymen*
D Company	14 tanks (M1)
E Company	6 antitank vehicles (ITVs) (1 platoon detached to another battalion)

Each company has additionally:
- 1 armored personnel carrier**: Communications
- 1 armored personnel carrier: Maintenance
- 2 armored personnel carriers: Medic
- E Company has all the above plus 1 armored personnel carrier: Command and Control

Scout Platoon	6 Bradleys (M3): Each M3 can put a reconnaissance team on the ground.
Mortar Platoon	6 armored personnel carriers with 4.2" mortar (also 2 fire direction tracks, 2 command and control personnel carriers)
Air Defense Platoon	4 Vulcan (antiaircraft tracks)
Engineer Platoon	4 armored personnel carriers

Battalion commander, battalion S-3, company commands all have combat vehicles.
Over 100 wheeled vehicles are in the battalion.

Enemy Defending:
Reinforced company
- 3 BMP platoons
- 1 or 2 tank platoons
- mortar platoon
- AT platoon

---

* Infantrymen referred to here are dismounted foot soldiers. Three additional men stay with each M2 Bradley fighting vehicle. Each dismounted platoon consists of 3 six-man squads and a platoon leader with his RTO, totaling 20 men. The company contains 3 of these platoons.

** Armored personnel carriers have the 50-caliber MG mounted on the cupola.

done that afternoon and night would further clarify each unit's part in the operation. Always echoed the lament of his staff that higher headquarters had not generated an adequate intelligence picture for detailed planning for the attack. The only bright spot during the briefing was the radio call from the scouts that they had completed their route reconnaissance and set up the designated checkpoints. All was in place for the night movement.

By 1445 Always realized that any further delay of the orders group would be counterproductive. He dismissed them with the following general orders: B Company, leading with its dismounted infantry, would spearhead the main attack along Axis RED followed by its own Bradleys and then C Company's tanks. A Company, also leading with dismounted infantry, would conduct a feint along Axis WHITE. D Company would hesitate, then commit along whichever axis proved most vulnerable. All committed forces would orient on Objective BLUE. E Company's antitank systems would provide covering fire from the high ground south of Checkpoint 2 (CP 2), as would the mortars. The scouts would secure the line of departure (LD) along the road running northwest to southeast below CP 2. Almost as an afterthought, having listened to the discussion of his S-2 and chemical officer, Always arranged for a smoke screen to be set down along that same road.

Even as they drove away, Always was having serious misgivings. If only he knew more about the enemy. If only he had more time. Why was it that his air defense platoon leader and engineer officer looked so uncertain? And the observers—they had certainly looked smug throughout the dreary briefing. He took a long pull on his canteen; the temperature on the desert floor had soared to over 105 degrees. Under the canvas cover of the tactical operations center, where all the orders group and observers had jammed tightly together, it must have been over 110 degrees. If the air had not been so hellishly dry they surely would have all been drenched in sweat. Instead white salt stains had broken out on their battle dress uniforms. Always forced himself to concentrate. He would have to get centralized control of this horde he was about to unleash.

Time continued to speed away from the colonel. Within an hour and a half of the order the quartering parties were readying themselves for departure at dusk. Always was surprised at both the size of the parties (one vehicle per platoon) and the necessity to get them out so far in advance. He realized how little time he had given his subordinate commanders to do their own planning and get their elements in motion.

He moved his command to his Bradley fighting vehicle. With darkness arriving soon he wanted to go over the procedures he would employ in order to both retain command and be prepared for personal combat from his Bradley. As he made the vehicle ready, his fire support officer came by to ask where he should be during the mission. This was a question Always had not considered before. After a quick discussion he decided to leave the artillery captain in the TOC. After all, that was where he kept his computers and the majority of his communications. It was also where he would best be able to keep abreast of the developing situation, with the intelligence officer present as well as the assistant operations officer. Always had already resolved that he was going forward with the main attack. It had long been his nature to go where the action was the heaviest, to lead by example, and to be able to make key decisions at the critical point.

Because of that decision he was anxious to move toward B Company during the last light of the day. Unfamiliar with vehicular movement in the dark, he wanted to get within sight of the company before the sun set and the units began to move out. He had delegated overall control of the battalion during the night road march to the operations officer, who was similarly situated in a Bradley. The executive officer would be fully occupied with the movement of the support elements that would have to refuel the vehicles at the conclusion of the road march and before the dawn attack, as well as recover and repair any equipment that broke down during the night. Again Always had been impressed with Major Walters' grasp of detail. He had covered every contingency, to include a quick resupply of ammunition and enough water to get the task force through the first day's fighting. In the desert, water might be the most critical supply element of all.

As his Bradley rumbled by the various formations, Always was inundated with diesel fumes. Thousands of tons of steel were readying for movement, and the massive engines it took to propel them forward filled the desert air with the heavy smell of exhaust. Adrenaline pumped into the commander's veins as the fuel of the two hundred vessels pumped into their engines. The ground seemed to vibrate as the monsters lumbered into position, tanks weighing in at more than sixty tons each, the infantry vehicles at more than twenty-five. It was exciting, and he was filled with an exuberance he had never felt with his light forces. It was as if he had an unstoppable force, impervious to anything that might be thrown at it. He was a giant, and each of these incredible machines was an extension of his sinews. Always was ready for battle.

Night fell as suddenly as morning had dawned. The stars came out and filled the desert sky. Mountain peaks faded in the distance, and the prominences Always had studiously memorized before dark now became confused in the uncertain, shadowed outlines against the black sky. The radio stayed silent. The plan was to move without any broadcasting, the better to delude the enemy, who surely must be scanning the frequency spectrum to pick up any chatter. Although the radios were secure, it was always safer to minimize traffic. Even if conversations could not be monitored, the mere breaking of the radio waves could tip off an experienced listener that something was about to happen. With the outside sounds muffled by his radio helmet (called a CVC), Always sat in the commander's hatch in relative solitude. Two feet to his left sat his gunner, eyes affixed to the thermal sight, peering out into the darkness. Several feet away, separated from the vehicle commander and his gunner by several tons of steel, sat the driver looking through a light intensification scope. Gunner and driver were both silent, allowing the battalion commander the quiet of his thoughts. The three of them could talk to each other over the intercom system, but for the moment Always preferred to reflect on the mission ahead. The peacefulness of the moment relaxed him, perhaps for the first time that day. Then Bravo Company moved out.

"Let's go." Always ordered his driver, Private First Class Spivey, to move.

The road march was uneventful. Occasionally there was a halt in the column as some congestion up front out of sight created a delay, but soon enough it was sorted out and the units resumed their movement. Always realized that he had little control over what was taking place. In reality he was just along for the ride. With no radio communication, with visibility reduced to a few meters (the colonel was not using any night vision devices, trying to adjust his eyes to the darkness and keep abreast of the movement by reading the terrain features he passed by), and with no contact with any of his soldiers save the two with him in his vehicle, Always could only assume from his own proximate location and the lack of calls on the airwaves that things were going more or less according to plan. As they passed an occasional scout checkpoint he congratulated himself on his forethought in placing them to keep everyone on track. Yet, not being sure of the condition of the rest of the task force caused a certain uneasiness in the commander. After all, they had traveled several dozen kilometers, and even at the outset the column, with all of its planned intervals between units and vehicles within the units, had been stretched out over more than a few kilometers.

Shortly before midnight Bravo Company arrived in its assembly area. The quartering party came out to meet it, each platoon representative picking out his platoon in the dark and leading it into predesignated positions. Movements were silent and efficient. Captain Baker had a crack company. Immediately it set in a hasty defense, sighting vehicle guns, disembarking infantrymen who began to dig foxhole positions out in front of the Bradleys, laying wire back to the company command post, and sending out a patrol to secure the outlying area from enemy probes. It occurred to the colonel that by organizing the companies as pure infantry or pure armor he had severely handicapped his tankers. They could not spare anyone from their limited four-men tank crews to flesh out their defenses. They would have to rely on whatever work the quartering parties had done to secure their positions. Patrols would be out of the question. He made a mental note not to repeat this mistake.

Always remembered what disdain he had held for mounted soldiers, whom he saw as essentially sheltered from the strain of having to march by foot and carry everything on their backs. These men had been working nonstop ever since he had arrived on the scene two days ago. Before dusk they had crawled into the back of their tightly packed vehicles, simmering in the late afternoon sun, then further heated by the crush of elbow-to-elbow bodies and running engines. They had been jostled over a four-hour road march and were now setting in an arduous defense, and facing an operation that would have them on their way by 0300 in order to get into position for the attack by dawn shortly after 0400.

Neither did the vehicle crews get any rest. They refueled their vehicles, checked maintenance, and manned their sights. There was no guarantee that the enemy was not lying in wait for them even within the assembly area.

Always realized again that he had blundered by positioning the scouts along the route without stressing that they clear the assembly area first. It was only sheer luck that the quartering parties had not run into a buzz saw of enemy. There were plenty of places for an ambushing force to hide. They might be there even now. Always wished he had kept the artillery officer with him. He wanted to develop a fire support plan, not just for the attack, but for defense of the assembly area as well. His units were vulnerable, and the thought of it accentuated the chill in the night air.

Concern overwhelmed him as he considered his lack of forethought. Here he was alone in the dark, aware of the status of only one of his companies, out of contact with his tactical operations center and under a self-imposed radio silence. In a few hours dawn would come. Would everybody be ready to kick off on time? What intelligence did they have of the enemy? Had Brigade passed them anything of use? He knew he was missing something there, he just didn't know what. Nor did he feel like calling on the Brigade frequency; he might reveal his own ignorance.

What if enemy air appeared at first light? He had not thought to designate priorities to his air defense platoon. He didn't even know where they were. Had they broken down to cover the various units? Were they covering the line of departure? What early warning system could he depend on?

My God, Always thought to himself, there must be a million things I forgot to cover. He felt miserable.

At that moment he sensed a dark figure climbing up on the deck of his Bradley.

"Excuse me, sir." It was Captain Johnson, his assistant operations officer. "I'm sorry it took me so long to find you, sir. I knew you were in Bravo Company's area, but I had to take my time to find you without compromising security. I have the status report for you."

Always tried to contain his relief, receiving the report as if he had expected it all along. The news was reassuring. Almost all the vehicles had made it in to their proper locations, and those that had not had been policed up by Major Walters bringing up the rear. He would have them rejoined with their units shortly. The air defense platoon was intact, confused as to what their mission was, but at least all accounted for and awaiting orders. Always told the captain to put two gun systems with each of the attacking companies. The teams of shoulder-fired antiaircraft missiles had already been attached to the combat companies and had been following the respective commanders in their jeeps.

"Have we gotten any intelligence about Objective BLUE?"

"No, sir." There was still a hole a mile wide there.

"OK, that's not great news, but we'll keep the attack on schedule, relying on our dismounted infantry to send us back reports as they make their way forward. Have the radio net opened up at 0330 with a radio check. We'll maintain radio silence until then, short of any emergencies. Tell Major Rogers to move up with Alpha Company and keep me advised on the progress they make on that flank."

"All right, sir. I'll pass all of that on."

"Thanks. Oh by the way, Captain Johnson, good job finding me in the dark. You've been a big help." Even though Always was embarrassed at his lack of forethought, he was a big enough commander to give credit where it was due.

As Johnson slipped away, Always looked at his watch. It was 0245. Bravo Company's infantrymen were moving off in the dark. Always was amazed at how few they numbered, three platoons of twenty men each.

In his perch in the Bradley, Always shut his eyes to doze. There was no point in fretting any more. The die was cast, and in a short while there would be enough new worries with which to preoccupy himself. Better to rest while he could.

He awoke to Sergeant Kelso, his gunner, gently nudging his elbow. "Sir, you're being called on the radio."

With a thick tongue he answered to his call sign, then waited while all the parties came up on the net call. It took longer than he expected. There were a lot of subscribers on the net, more than he had imagined, and some of them had to be called several times before they answered. The air force liaison officer never did answer, nor did the two dismounted companies moving forward at this time. Their portable radios were operating in the clear and could not hear the encoded call. This created a dilemma. Having a mixture of secure and unsecure parties on the net had the effect of talking on two nets on one frequency. The resolution was to operate totally in the clear, with the ensuing risks to security. Always decided that he would preserve the secure net a little while longer. He hoped the dismounted infantrymen would have the sense to call in should they discover something critical or get involved in a heavy action. Captain Baker had gone forward with his infantry, while Captain Archer had remained with his Bradleys, letting the senior platoon leader take charge of the dismounted formation. Always wasn't sure which one was right. At the moment, however, one of his commanders was off the net. That and the missing air force officer worried him.

The scouts were dutifully in position. They reported they could see little from where they were on the line of departure, but that they had passed forward both dismounted companies in the last thirty minutes. So far there had been no incidents. Always confirmed the plan. By the time the conversation was completed it was 0350 and there was an almost imperceptible lightening of the horizon in the east. It was time to move. In fact, it was clear that they could not now hit the line of departure on time. Too many precious moments had been spent in establishing the radio net.

Even as they began to move forward—B and C along Axis RED, D along Axis WHITE—the task force commander began to lose control. Precisely at 0400 the preplanned smoke began to fall across the front, but the wind had now shifted to a northeasterly direction and blew it back over the moving vehicles, making their journey difficult. Although they were only twenty meters away, Always could not see the nearest vehicles of Bravo Company, forcing him to proceed by dead reckoning in what he thought was the general direction of the attack.

At 0410 he heard Major Rogers call in the crossing of the line of departure (LD) on the left, and a few minutes later he saw the road that he had designated as this key control measure. But he had not yet heard from the infantry. He was still attacking blindly. Moreover, to continue moving forward with the armored vehicles risked running into his own dismounts. In the smoke and the poor light, with nervous trigger fingers itching for combat, the chance of shooting his own men, of fratricide, was extremely high.

"Oscar 42, Poppa 42, pull into some cover and establish contact with your infantry. Give me a situation report before you proceed." Always called to Bravo and Alpha companies.

In a few minutes A Company reported that it was still about 1,500 meters short of Checkpoint 1. Captain Baker, on the ground with his dismounted soldiers, called directly to Always, who had to switch to unsecure on his radio to answer him. B Company's dismounts were nearing CP 2, but it appeared as if an obstacle was located there; they would have to move cautiously. Always heard artillery fire coming in over to his left in the vicinity of Echo Company. He terminated his conversation with Captain Baker and shifted back to secure in order to find out what was going on. He had not heard any of the conversations of the last few minutes because when he had left the secure mode to converse with Baker he had essentially left his own net. Each shift of mode on the radio necessitated a contortion by Always in his seat. The radios were located directly behind the small of his back; in the narrow confines of the commander's cupola he had to drop down, swivel around, find the switch in the gloom of the vehicle interior, and change its position.

As he came up in secure he heard Captain Evans talking to the TOC. ". . . intensive fire coming in. I'm taking casualties. The enemy has a fix on my location. I've got to pull out of here or get chewed to pieces."

Always approved the move, but neither battalion nor company commander was sure where Echo was going to go. To move forward would put it out in front of the attack. To move back would put it out of the direct fire range of its own guns, unable to reach to the suspected enemy positions. The precision of the incoming artillery indicated that the enemy had placed his own reconnaissance in position the night before, probably as the first elements of Always' quartering parties were seen coming in. He had fixed E Company's exact position, and even in the smoke had been able to bring Evans under effective fire. A minute later the mortar platoon reported that it was also taking artillery fire. Two tubes had been destroyed in the first barrage, and several of the mortarmen wounded. The mortar platoon, too, would have to pull out.

Captain Baker was calling in the clear. "The obstacles are covered. I'm taking small arms fire and incoming mortars. This is going to take awhile."

Always needed some movement on his left. He attempted to call A Company's dismounts directly, but they were either out of the range of their portable radios or an intervening terrain feature was interrupting the transmission waves. The colonel had now been operating in the clear for two minutes and he needed to get back to secure.

This was the moment that Brigade chose to call for a situation report. Even as he answered he found himself taking incoming artillery fire. He reached up to close the hatch cover as he responded to the call from higher headquarters. Simultaneously E Company reported in its new position as the TOC overrode Evans' call to yell out that they were under artillery fire and would have to move. The noise had become deafening—the artillery fire, the roar of the Bradley engine as Spivey pulled them out of the fire, the chatter on both radio nets, and the conversations between the colonel and his crew on the intercom as the battalion commander tried to keep himself moving with Bravo Company. The artillery fire was eating them alive. He should have swept the area for enemy scouts. A mere one or two of them, situated close enough to observe the task force's movements in and around the LD and the assembly area, was throwing the entire attack off balance.

Brigade was not happy with the report and encouraged Always to get moving. Now Alpha Company was calling for artillery support as it neared CP 1. It had picked up a report of Bravo's fight at CP 2 and was proceeding with caution. The artillery was now having to choose between supporting Bravo and Alpha or concentrating on countermortar or counterartillery fire to ease the dilemma of the task force. Always was hard pressed to give guidance, not because he did not know what he wanted to prioritize (it was crucial that B Company break through the obstacle) but because his artillery officer was caught in the dislocation of the TOC and was momentarily off the net. That left him only the mortars to turn to, but they were down one third of their guns, and the other four were on the move.

The key seemed to be B Company. If it could get through the obstacle, the task force could ram home the main attack.

"Oscar 42, this is Lima 42, over." Always was trying to reach Captain Baker.

"Oscar 42, this is Lima 42."

No response.

"Oscar, this is Lima."

The colonel's voice was hardening. He remembered Baker was operating without secure, cursed himself, and flipped the radio switch. "Oscar 42, this is Lima 42."

Heavy breathing. "This is Oscar 42, over."

"This is Lima. Sitrep."

"Roger. We're closing in on the obstacle slowly but surely. I've taken seven casualties, three killed, four wounded. I'll need to get them out or two of them will be done for. Over."

"Oscar, this is Lima. The whole shebang is dependent on you getting through that obstacle. That's got to be your first priority. Put a full court press on, and get through that obstacle. Call me the second you make it. Do you roger?"

"This is Oscar. I roger."

Always' mind was racing. He could not be sure how long Baker would take and he had to get something going. The whole attack was falling apart at the line of departure. He considered committing Charlie Company to try to penetrate at CP 4, then dismissed the idea. There was still a chance to put the plan into effect if Alpha could close in on CP 1.

"Lima 42, this is Lima 51, over." It was the smoke platoon.

"This is Lima 42, over."

"This is 51. We will be out of smoke in five minutes."

"This is 42. Roger, out."

"Poppa 42, this is Lima 42."

"This is Poppa, over." Alpha Company answered quickly.

"This is Lima. I need you to launch your vehicles up to CP 1. Try to coordinate it with your ground element. We're being held up in the south, and I need you to punch through."

"Wilco, over."

"This is Lima, let me know when you get there. Out."

The traffic on the radio was picking up speed. At any given moment three or four stations were trying to reach Always. The TOC was now back in operation and the fire support officer was seeking guidance. The air force liaison officer entered the net, unsecure—yet another station necessitating a quick switch off the secure mode. Charlie Company came under artillery fire and had to shift position. Captain Carter was eager to kick off his attack, but Always held him back until he could get a clearer picture. Brigade called two more times, the demand in Always' commander's voice quickening.

Five interminable minutes passed. Then, "Lima, this is Oscar, I'm through the obstacle."

Always' heart jumped. "This is Lima. Say again, over."

"This is Oscar. I say again, I'm through the obstacle."

In a flash Always acknowledged the call, signaled the vehicles in Bravo to move forward, and directed Charlie to follow. Just as the smoke lifted they dashed across the tank trail straight at Checkpoint 2, twenty-three vehicles still able to move after the intense artillery pounding in the two companies. Missing were the air defense teams mounted in the jeeps. Without armor protection they had been killed in the initial barrages.

Suddenly Bravo Company lurched to a sudden stop, forcing Private First Class Spivey to scoot his vehicle to the right to avoid crashing into its tail vehicle. Always smashed his face into the sight to his front. Buttoned up in the vehicle, with his hands grasping the map and the various communications controls, shifting rapidly to the turret controls whenever he sensed the likelihood of enemy engagement, he was careening back and forth in his seat like a top out of control. A trickle of blood worked its way down his nose.

"What in the hell is going on?"

His question drifted over the intercom system, bringing a polite response from his gunner, "I don't know, sir."

Dust flew every which way, obscuring vision worse than had the smoke. Always reached up and popped his hatch. He had to risk a look through the open hatch. He was completely blind at the moment through any of the periscopes or vision devices.

The scene that came into view seared his eyes. Stretched across the narrow pass at CP 2 was an obstacle of concertina wire, mines, and a four-foot-deep tank ditch. On the far side was Bravo Company's infantry, gone to ground in a firefight with the enemy defenders 100 meters farther up the draw. And beyond that were the enemy T-72s and BMPs, dug in up to their gun tubes so that it was almost impossible to see them except when they fired, and virtually immune from any direct fire. And at that moment they were picking apart the Bradleys of Bravo Company.

Captain Carter had already reacted and was desperately attempting to pull his tanks out of direct fire of the T-72s. It was a wild confusion of pivoting tanks and exploding Bradleys. A few of the vehicles intentionally poured out covering smoke from their diesel fuel to make good their escape, adding to the confusion and saving themselves, at least for the moment.

Always yelled to Spivey to pull back, and fired off one burst of the 25mm gun at a BMP firing at him from 500 meters up the draw. Both guns missed, and Spivey made it back around the bend, hugging the side of the cliff on the southern extremity of Hill 826.

In an instant Always reached the scout platoon and directed it to move to CP 4 to see if there was a way through to Objective BLUE from that direction. Prompted by his TOC, Always remembered the engineer platoon, hitherto forgotten, and ordered it to fall in behind the scouts to reduce any obstacles they might find. Major Rogers called on the radio with a situation report on CP 1.

With a little luck A Company had linked up with its dismounts and been able to close on the obstacle blocking the valley at CP 1. Although it had taken a few casualties and had lost one vehicle, Alpha was making progress on reducing the obstacle and would be through it in another few minutes. The bad news was that there were at least two more obstacles several hundred meters behind the first, and each of them was covered by fire. Progress was possible, but it would be slow and probably come at some expense.

Always had to make a decision. Neither option—an attack

through CP 1 or one through CP 4—was a sure thing. One thing was sure, though—an attack through CP 2 was out of the question. It was a kill zone that had already decimated one company. He could ill afford to risk another in the attempt. Yet he still knew nothing of CP 4. His inclination was to mass forces, but by putting all of his eggs in one basket he risked abject failure. His mission was to take Objective BLUE, and he had to take it, even if it meant heavy losses. On that note he hedged his bet. Charlie Company and the approximately one remaining mounted platoon of Bravo would move in behind the scouts and engineers at CP 4. Bravo's dismounts would continue to work up the draw north of CP 2. Delta Company would reinforce A Company as they worked into the valley beyond CP 1. Artillery priority went to Alpha, where there was known enemy. Mortars would give what support they could to Baker and his infantry. Always would move up with D Company to the northern effort. Carter had command of the effort at CP 4.

It took Always about ten minutes to pass the requisite orders, complicated by his having to shift back and forth to secure. He stifled his anger at Baker, who had implied the obstacle at CP 2 was down, but resolved to make it clear later that whether or not ground troops had personally passed through an obstacle was irrelevant. It only mattered that they had reduced the obstacle so that the heavy forces could follow them through. That had been a momentous blunder, but Always realized that he himself was not above reproach in confusing the issue.

The fight now turned into a battle of attrition. By the time the task force commander had made it up to CP 1, the scouts and engineers had found the expected obstacle at CP 4. It took them ten minutes to reduce it, and in the process more than 50 percent of the engineer platoon was killed or wounded. The scouts who picked up the mission of clearing the enemy infantry in and around the obstacle also suffered heavily; thereafter, they lost their capacity to dismount. Once the obstacle was clear, the tanks and Bradleys under Carter could advance only with great care; dug-in infantry armed with antitank weapons guarded the slopes of both sides of the narrow valleys and had to be eliminated methodically before the armor could proceed. Without dismounted infantry this became an excruciatingly slow process. It was two hours before Objective BLUE was reached from that direction.

In the center, Baker brought up the majority of his infantry, moving cautiously and without adequate fire support. The air force liaison officer, who had missed the orders passed by Always over the radio (throughout the fight he was alternately on and off the net), brought in an air strike about 0830 along what he still believed to be the main axis of advance from CP 2 to CP 3. He had heard the fighting from his position in the vicinity of Hill 826, which he had valiantly climbed under fire, but assumed that U.S. troops were buttoned up inside their armor-protected vehicles. This assumption cost Baker 20 percent of the remainder of his infantrymen, lost to friendly air, although it also inflicted heavy damage on the enemy infantry. By this time, with no armored threat along this approach, the enemy had shifted the majority of his tanks and BMPs back to Objective BLUE. Baker's progress was slowed by his heavy casualties, but he pressed on to the objective, gradually linking up with Carter southeast of CP 3.

The main attack of Alpha and Delta companies became a drawn-out battle through the pass south of Hill 785. Blinded by an approach that took them directly into the morning sun, the attackers suffered a major disadvantage. Nonetheless, aggressive tactics and good gunnery brought the exchange ratio about even. A major setback occurred, however, when enemy air struck at about 0900. In the confusion of the morning's movements, the air defense platoon, which had never been directly attached to any specific company and, therefore, had failed to put its radios on any commander's net save that of the task force, had

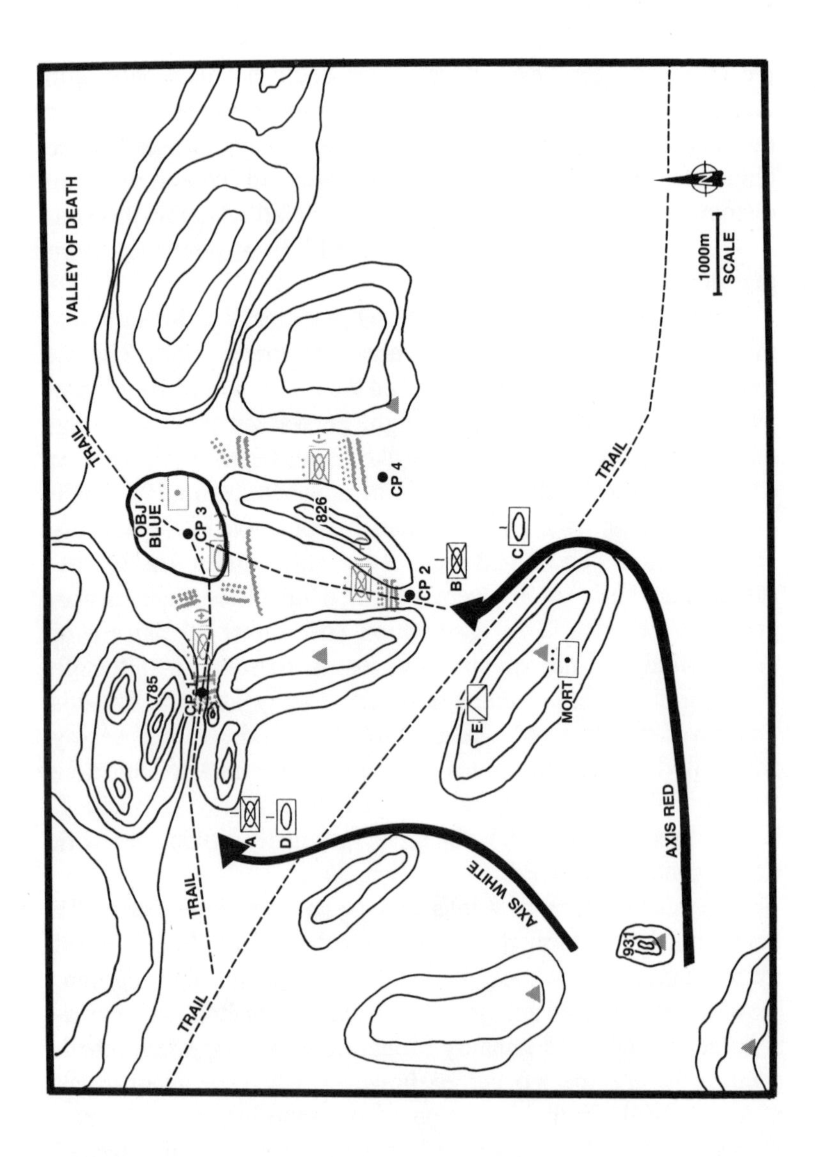
VALLEY OF DEATH
TRAIL
OBJ BLUE
CP 3
CP 4
826
CP 2
B
C
CP 1
785
A
D
E
MORT
AXIS RED
AXIS WHITE
TRAIL
TRAIL
TRAIL
931
1000m
SCALE

not kept up with the forward elements of the attack. With the vehicle-mounted antiaircraft guns too far back, and with the missile teams killed in the dawn artillery barrage, the enemy air force had virtually free rein over the main attack. Only the Bradleys, with their superior tracking ability, were able to threaten the fast movers. But all they did was threaten, and five tanks and three Bradleys were destroyed before the enemy ran out of ammunition and pulled away.

By 1000 Always' task force had converged on the edges of Objective BLUE, driving back the enemy before its massed fire-power. By this time only C Company was relatively intact. E Company was the next best off, but Always, in his confusion, had neglected to give explicit orders to Evans. He therefore elected to follow the main attack over Axis WHITE and through CP 1. Here the terrain restricted his ability to bring fire on the enemy's armor. By the time he did deploy on Objective BLUE, the enemy had decided to make their escape to the northeast.

By 1045 Always could report in to Brigade that he had taken Objective BLUE and was reconsolidating for a possible enemy counterattack. More than 50 percent of his combat vehicles and 80 percent of his infantry had been knocked out of action. His engineer platoon had been annihilated, and his scouts were in rough shape. A resolute counterattack would have been hard to defeat, and at the moment Always was hardly in a position to continue attacking. His medical support was severely over-stretched, and two medic tracks had been knocked out in the fighting. Only his combat service support was in good shape, husbanded with great care by his XO, who brought it forward as soon as the ground was secure in order to refuel and rearm the combat forces.

As the medic put seven stitches in the gash over Always' eye, the colonel reflected on the mess he had made of things. Just then Lieutenant Colonel Drivon, the evaluator, drove up in his jeep.

"Well A. Tack, how do you feel?" This was obviously Drivon's best effort at being friendly, addressing Always by his first name.

"I feel fine, thank you. How are you?" Always was not about to admit anything.

Ignoring the question Drivon issued instructions for the time and place of the after-action review of the operation. All of the task force commanders and principal staff would report to a designated grid location at 1230. The battalion could expect to receive their next mission within the hour. Only those forces that could be reconstituted would be available for the next mission.

"What about my dead and wounded?" Always asked. "After all, they were dead to begin with."

"If you report your dead and request replacements, we resurrect them at midnight. If you treat and evacuate your wounded, we heal them and send them back to the replacement system. It's up to you to get them back to the front. The equipment works more or less the same way. If a tank or Bradley has been blown apart, we'll see you get replacements if you work the system. If it just needs repairs, then you've got to do it. No free lunch here, you know. Little bumps and bruises, like your face there—well, you just have to live with that, so to speak." Drivon cleverly concealed any sympathy he might have for Always.

"Okay. Thanks. I'll get my people up to you at 1230."

The next few hours were hectic. Always called a hasty meeting with his S-3 and his XO to arrange for the reconstitution of the battalion. The order to continue the attack to the northeast the following morning arrived around 1200. Always had time only to look at the map, give some very general instructions, and pass on a warning order to the companies. By that time he had to leave for the review session along with his entire staff and all his commanders. The assistant S-3 was left in charge of the planning. At every level assistants would have to do yeoman work to get the battalion back on its feet in time for the morning attack.

The observers had gone to great lengths to make the review site difficult to find, putting it deep in a ravine. But the task force leaders found it, squeezed their tired and smelly bodies into the briefing van, and listened to what the observers had to say. Although the graphic descriptions of the errors stung, they were in every case accurate. At appropriate moments particularly glaring errors were played back on voice recordings and videotapes. There was a great shot of Always' face, bleeding and dazed, peering out of his Bradley into the destruction of B Company at CP 2, followed by the tape of Captain Baker reporting that he was through the obstacle. No one chuckled.

When it was over, some two excruciating hours later, Lieutenant Colonel Drivon and his team left the van at the disposal of Always and his men. The commander took advantage of the opportunity to make a little speech to his men, a speech that avoided apology or accusation but did not deny failure. He praised the commanders and their men for their resoluteness in the face of the enemy, and rededicated their mutual effort to figuring out where they went wrong and putting it right. With that done he dismissed the group and headed back toward his headquarters. It would be dark in four hours, and there was much work to be done.

On the drive back Always mulled over the lessons he had learned during the preceding twenty-four hours:

Intelligence is the building block of the order. Don't expect it from higher headquarters. Build it from the bottom up, and make it specific. Get the scouts out early, and leave them where they can tell the commander what's going on. At the same time, defeat the enemy reconnaissance effort. Blind him. Don't let him get a fix on you.

Get the special staff under control. The artillery officer

and the air liaison officer are too critical to let them get out of touch. Put them in your pocket and keep them there. Have the chemical officer talk to the intelligence officer. If the winds are not right, don't use smoke. If you can blind the enemy and retain your own freedom of movement, use all the smoke you can get.

Use every available minute to get things focused on the central mission. Alert the subordinate elements as soon as you have an indication of the upcoming mission. Get the staff in motion quickly, but keep them on track. Update as intelligence gives you a better picture. Don't worry about picture-perfect written orders. What orders you do give, make clear. Above all, get your intention across to every key subordinate, then make sure they get it passed on to their subordinates, and so on down the line. When you give the order, do so at a place and in a manner that facilitates understanding of the mission. Key ground overlooking the battle area is probably the best place to do that. Leave your subordinates plenty of time to do their own reconnaissance and planning.

Put yourself at the critical place. A picture is worth a thousand words. If you can't be at the critical place, make sure you understand what the guy there is telling you. Ask the critical questions. Don't be pressured into making a hasty decision, but don't equivocate either. When you do make a major shift, ensure everybody gets the word. Don't assume that they will. Hold someone accountable for informing them.

Take care of the engineers and the air defense units. Husband them, protect them, and give them clear and specific orders. When it suits the mission, put them under control of a subordinate commander, then hold him accountable for them. When you have to retain them under task force control, have them under tight hold, but don't stifle their initiative. Get the air defense missile teams under some armor protec-

tion, probably right with the company commander they're protecting.

Coordinate your combat power. Bring the mass of the tanks and Bradleys together at the point of main effort. Coordinate that effort with the infantrymen. Cover them with artillery and mortar fire. Don't give the enemy room to escape. Cut him off, overrun him, and annihilate him.

Get the radio net under control. Everybody has got to be in instant communication with everybody else. Get the commanders to talk to each other, over the task force net if no other way is possible.

The list went on, from the grand to the small, but at that point the jeep pulled into the TOC set up on Objective BLUE. It was time for Always to turn his attention to the next mission. His jeep driver gave him a wide smile and a salute as the colonel stepped out. "There's a good soldier," the commander thought to himself.

CHAPTER 3

Change of Mission

The morning's attack had the task force moving more than twenty kilometers to defeat the enemy in zone and secure Hill 781. Reportedly, the enemy they had knocked off of Objective BLUE had fallen back and been reinforced, so that two motorized infantry companies reinforced with one or two platoons of tanks were now digging in around Hill 781. This would comprise a larger force than they had faced that morning, which was now estimated to have been a reinforced company. (It was with some chagrin that Always accepted the fact that one company had done so much damage to his task force.) By attacking at first light there was a chance that they could catch them before they could improve their defenses.

Despite the fact that the scouts were not yet reconstituted, it was imperative that they get out forward before much more time passed. The distance implied a great deal of risk, and Always was not about to attack blindly a second time. As more scouts replaced the casualties from the morning, they could link up forward and flesh out the intelligence picture. The enemy could be anywhere between BLUE and Hill 781, and Always was not about to be caught unaware. The plan he would develop now would be modified as more information came in. If he had enough infantry reconstituted he would send them out as well, but since their ranks were so thin he would hold them back and use them in the morning attack. Always was careful to brief the scout platoon leader, Lieutenant Wise, an immense young man brimming with energy and zeal, picked for his strong leadership abilities and keen intelligence.

Major Rogers had brought together the entire staff, receiving guidance and giving estimates to the commander. The give and take was much more open this time, Always stating what he felt was essential to the proper accomplishment of the mission, Rogers and the staff formulating alternatives and estimates out of those essentials. It was an efficient way to develop an early plan, but it was not without its risks. The intelligence picture was still completely bare. By pulling in the air defense officer and the engineer platoon leader, their respective units were left without their leadership for the amount of time they spent with the staff. Both those young lieutenants were torn two ways, staff and command. They had had no rest in days, and it was clear they were not going to be getting any soon. When they did complete the planning, which would not be until late, they would have to hasten back to their platoons to get them ready for the execution. They depended on the strength of their platoon sergeants to keep things going until they got back.

To help the battalion with its difficult mission, Lieutenant Colonel Always would receive the support of an attack helicopter battalion. This was a major addition of combat power. Major Rogers raised the delicate issue of the need to coordinate the aviation activities with the artillery, air force, and air defense plans. There was a major opportunity here for error, with one getting in the way of the other, or worse, inflicting casualties upon the other. Always gave Rogers stern instructions to work this out with all parties. A liaison officer from the aviation unit was due at the TOC shortly before dusk.

Major Walters joined the staff midway through the discussions and reported that reconstitution was moving along smoothly. A herculean effort was being put forth to repair the equipment. Always gave some guidance, prioritizing the tanks and air defense guns for repair, although it was imperative that enough Bradleys be brought into the battle by dawn. Somehow Walters had been able to get a hot meal prepared, which would be coming up just before dark. That would help out, as food would compensate to some extent for the fatigue the men were now feeling.

Although time was slipping away rapidly, Always felt it necessary to get a look at the route of movement as far forward as it was safe to do so. Since yesterday's reconnaissance from the air had proven deceptive to actual conditions on the ground, he elected to take his Bradley this time and get a closer view. He would have to be careful not to fall prey to an enemy ambush, but the risk was worth the look. He took the operations officer's Bradley with him for security, having one of the sergeants from the TOC take the place of Major Rogers so that he would be free to continue planning. As it turned out, this little trip went a great deal toward giving Always a feel for the early conduct of the operation.

The operations order was given at 1730, the orders group assembling on the high ground overlooking the valley where they would attack in the early stages of the operation. A few spot reports had come in from the scouts, and one of the infantry patrols clearing the close-in area around BLUE had picked up a prisoner, an enemy scout left behind to watch and report on the task force. He was recalcitrant, but a map on him gave a clue, nothing more than that, of some enemy dispositions in the valley below. It was enough to divert the few scouts Always did have to check out the possibility of an ambush beyond CP 2.

The order went much smoother this time. The written product was reduced to only two pages, the bulk of the information being written on the operations overlay prepared for each subordinate element. A great deal of the administrative instructions that had dominated the last order was omitted from the briefing and placed in a written annex to be given to the company execs. Only information that was pertinent to the operational mission was included in the order. Nonetheless, there were still a few vital holes. The aviation coordination still had not taken place, so their incorporation in the plan could only be tentative. The helicopter liaison officer did arrive toward the end of the briefing, but it was too late to confirm exactly how the aviation assets would be employed. The best employment of the engineers could not be determined until more intelligence came in. It was clear that their priority would go to the mobility of the attacking task force, but the objective was so far away, uncertainty remained as to terrain conditions and the location of enemy obstacles. Ominously, the intelligence officer briefed of the possibility of chemical release by the enemy. This necessitated the movement of the battalion in chemical suits, with masks at the ready for instant donning. This could take a heavy toll in a movement that would last several hours into the heat of the day.

At the end of the briefing, Lieutenant Colonel Always stood before his battalion's leaders and explained his intentions for successful accomplishment of this mission. He sensed their complete support and was quietly astonished at their willingness to place their faith in him again despite the rough mission he had led them on that morning. He acknowledged the uncertainties that lay before them, and ordered that a radio net call of all key subordinates take place at 0330 to share in the latest intelligence uncovered and to make any modifications necessary to the existing plans.

In the last ten minutes of the briefing Always drew out the questions of his subordinate commanders. He had gained an appreciation of the complexity of their jobs, and realized that any misunderstandings of intentions could result in confounding the entire mission. As soon as they left the TOC they would be racing to do their own planning, orders, and preparations. Any confusion would be compounded a hundredfold before dawn came. It was, therefore, with great care that Always listened to their comments, and only when he was sure that each man understood what his battalion commander wanted him to do did he dismiss the group. As he watched them go, he hoped that he had given them enough time to complete their preparations. It would be nine hours before the attack kicked off at 0400, but all of that time would be dark, and there was a myriad of tasks yet to accomplish.

Command Sergeant Major Hope had been present at the briefing, and in the short interlude after the meeting and before Always moved on to other matters, he came up to render a report on the state of the battalion. Morale was high, discipline holding; although the men were feeling fatigue from their exertion, they had plenty of starch left in them and were eager to get another crack at the enemy. As the noncommissioned officer spoke, Always realized how many problems Hope had taken care of for him during the day. Not that Hope claimed any credit for having done so; he was much too modest a man for that. But it was clear that he had taken a great deal of the burden off his commander by setting things right where they had gone astray, by understanding the intentions of the commanders and putting forces in motion to accomplish their ends. It was ironic that this soft-spoken, gentlemanly sergeant could breathe such fire into the men. Several times during the day, Always had seen an NCO fall afoul of the sergeant major by failing to follow up on responsibilities, only to receive a blistering admonishment that left him with a preference for contact with the enemy over another engagement with the wrath of the battalion's top soldier. Yet for the most part his demeanor was quiet, reassuring, offering encouragement to the soldiers to redouble their efforts, and offering praise for all they had done well. Only at the end of their discussion did Hope inquire as to the gash over

Always' eye, express concern that his commander take care of himself, and mildly suggest that he take an opportunity to get some sleep that night. Always marveled at the balance in his sergeant major—strength without arrogance, authority with deference, concern without solicitude. There was a lot of leadership in that man and he was glad he was on his side.

At 2000 Always moved out to his jeep in order to visit the subordinate units as they conducted their preparations.

"Good evening, Specialist Sharp. How are you doing?"

"Good evening, sir. Fine, thank you. Sir, I've saved you some supper. It's pretty good. Roast beef, corn bread, and some peaches. I got you some salad, and here's some salad dressing. I've got us a thermos of coffee. Do you take cream and sugar?" Sharp was a good soldier. He had stayed up on the radio all day, following the battle as closely as he could, eager to get into the fight, but prudent enough to wait for his commander's call before he came up. He had not missed a beat, ensuring the radios were set, recording the call signs and frequencies on a handy pocket card (called a "cheat sheet") for Always, updating the map, and now making sure he had a full meal for his commander.

"Thanks, Eric. This is great." It was the first time Always had addressed any of his men by their first name. He believed in formal address; it went with his strong sense of discipline. Yet it seemed somehow awkward in the face of Sharp's exuberance to remain so stiff. "Let's head on over to B Company." Always wanted a few quiet words with Baker, whose loose use of words had caused so much damage that morning.

"Did we kick their ass, sir? I mean did we beat the enemy good?" Sharp was eager for the commander's interpretation of events. He had had several conversations with some of the other drivers during the afternoon, and he wanted support in his high opinions of how well the battalion had done.

"Well, we kicked them off the objective. But to tell the

truth we paid too high a price for it. We shouldn't have let them get away either." Always was aware of how meaningful his comments would be, echoed a hundred times as they passed from mouth to mouth. "One thing you can be sure of—the men fought well, and the enemy sure as hell knows he's been in a fight with a top-notch outfit."

Sharp smiled as he pulled up in front of Captain Baker's command post. He had a good report to pass on to some of his buddies in B Company.

Always spent twenty minutes with Captain Baker, avoiding any harshness in his voice, sharing culpability in the morning's error, and encouraging him for a renewed effort the next day. Bravo Company would be split again, with some of their infantry being helicoptered in on the far objective at dawn, while the bulk of the force, the armored vehicles, would move to eliminate the suspected enemy waiting along the route (scouts were trying to confirm their disposition even as they spoke), and then culminate in the supporting attack by the helicopter-inserted infantry on 781. It was a tough mission, and evidence of Always' continued faith in Baker's abilities. The two commanders parted on an upbeat note, Always directing that Baker put himself with his mounted force this time, leaving the senior platoon leader to take the airmobile forces in.

As he moved around in the darkness of the night Always could sense the anticipation of the upcoming mission. A great amount of activity was taking place, from reloading ammunition, to maintaining equipment, to pockets of soldiers forming to receive hushed orders in the dark by the poncho-covered glow of red-filtered flashlights. Now and then Always stopped to talk to a soldier, sometimes with the soldier not discerning in the dark that he was talking to his battalion commander.

At 2300 Always listened in on the orders Captain Dilger gave his platoon leaders. His company had now been converted to what the military calls a "team" (implying combined arms

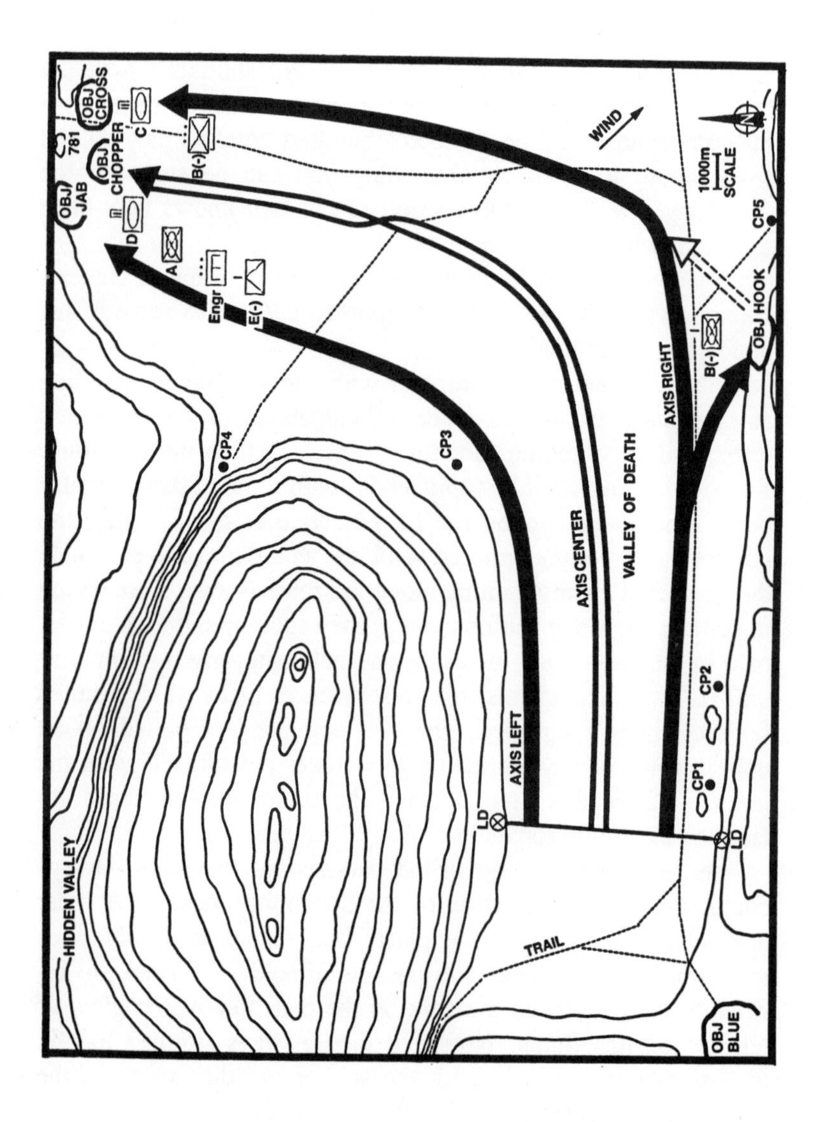
OBJ CROSS
781
OBJ CHOPPER
OBJ JAB
C
B(-)
D
A
Engr
E(-)
WIND
1000m
SCALE
CP5
OBJ HOOK
B(-)
AXIS RIGHT
VALLEY OF DEATH
AXIS CENTER
AXIS LEFT
CP4
CP3
CP2
CP1
LD
LD
TRAIL
OBJ BLUE
HIDDEN VALLEY

Map 2. Second Attack

Several false reports confuse the situation as task force moves to CP 4. Lieutenant Rodriguez's element is decimated. Eventually, motorized rifle regiment attacks by task force and continues to south.

OBJ CHOPPER:	2 infantry platoons inserted by helicopter (Lt. Rodriguez in command)
OBJ JAB:	Team Delta (2 tank platoons; 1 Bradley platoon with infantry)
	Team Alpha (2 Bradley platoons with infantry; 1 tank platoon)
	Company E (-) (1 ITV platoon)
OBJ CROSS:	Team Charlie (2 tank platoons, 1 Bradley platoon)
OBJ HOOK:	Team Bravo (2 Bradley platoons, 1 tank platoon, 1 ITV platoon, 1 infantry platoon)
	Team Bravo attacks HOOK; then CROSS, following Charlie.

Helicopter gunship battalion in support.
Engineer platoon provides mobility support to main effort on Axis LEFT.
Mortar platoon in general support (e.g., provides support on call to any company or team).

Note: One antitank (ITV) platoon detached to sister battalion.

at the company level), having lost one armor platoon and gained an infantry platoon. Dilger did a good job of specifying the tasks for his tanks, Bradleys, and infantrymen. He would have twenty of the latter, assuming all the replacements made it in at midnight. D Company had been given the lead in the main attack. Over a movement of this distance, with all the uncertainties that implied, it would be best to lead with tanks with enough infantry nearby to react to any unpleasant surprises the tanks could not deal with. The engineer platoon leader was also present. He would be following some distance behind Dilger, ready to react to the captain's call should Delta encounter obstacles. Satisfied that Dilger understood his intent properly, Always moved back to the TOC, but not before giving a word of praise and encouragement to the platoon leaders. He also stopped by to make a specific acknowledgment to the company first sergeant, whom Hope had singled out in his discussion earlier that evening for a particularly exemplary action.

Back at the TOC the staff was waiting for him with an update on the intelligence picture, a weather report, and recommendations on modifications of the mission. Major Rogers had done a superb job of integrating the efforts of the several staff officers, who with the exception of the aviation officer had completed their coordination down to the company level. The limitation on helicopters flying at night had forced the liaison officer from the aviation battalion to depart at dark; he, therefore, had missed some of the late-breaking intelligence. Despite this bothersome omission, Always was able to complete his guidance to the staff as the final order took form.

The task force would cross the line of departure at 0400 along two axes of advance. The main effort would be made in the north toward Objective JAB along Axis LEFT, Team Delta leading, followed by Team Alpha, then by the engineer platoon and Echo Company with only one platoon left under its command and control.

The supporting attack would proceed along Axis RIGHT, led by Team Charlie. Team Bravo, with only one platoon of infantrymen staying with it, would advance to Objective HOOK, where the scouts had located a platoon of BMPs waiting in ambush. They had already reduced an obstacle that was set at CP 2, but which the enemy had left uncovered. The scouts would leave an element there all night to ensure it was not reset. Bravo would additionally have a platoon from E Company attached to it. Once HOOK was reduced, Bravo would follow Team Charlie (two tank platoons, one platoon of Bradleys without dismounts) onto Objective CROSS to the east of Hill 781. The airmobile assault consisting of two platoons of infantry from Bravo would take off at 0415 and go into a landing zone (LZ) south of Objective CHOPPER. One scout team was making its way up there now to mark the LZ.

At the culminating point of the attack, pressure would be put on Hill 781 from the east, west, and south. Always was trying hard to achieve the mass he had missed the previous morning. To ensure that he got it the helicopters would approach along the high ground north and west of Axis LEFT, adding their fires on Objective JAB and then on Hill 781 at the crucial moments. Since the winds would be blowing from the northwest, smoke would be used to cover the crossing of the LD at 0400 and the approach of Bravo on HOOK. As the task force approached JAB and CROSS, the infantrymen on CHOPPER would fire up their smoke pots to screen their approach from direct fire. Artillery priority would go to Team D, mortar priority to Charlie. Since the pass at CP 4 was a possible enemy avenue of approach into the flank, an artillery-emplaced mine field would be planned northwest of the checkpoint, where the pass narrowed to single vehicle width. Air defense would be split left and right to cover both approaches. Always had been careful to give explicit instructions for the command and control of the air defense assets.

The operations officer would go with Team C to overlook the supporting attack. The fire support officer and the air force laison officer would come with Always in an armored personnel carrier following behind his Bradley along Axis LEFT, tucked in behind Delta. The colonel wanted to be sure he did not lose control of either critical fire support officer, although the air force officer protested that he would be severely restricted without his jeep to talk to the air force. After some severe squeezing by the infantry colonel, the young lieutenant admitted he had a portable radio, but protested it did not work very well. Always growled at his communications officer to fix the necessary radios to keep him in contact with his fire support. The whole business seemed excessively unwieldy to the task force commander.

At 0045 Always made his way over to his Bradley to get some sleep before the intelligence update at 0330. Spivey and Sharp had stretched out a cot between the infantry vehicle and the jeep. With a quiet word of thanks, and without taking off any of his uniform, Always collapsed on the cot and drifted into a deep sleep.

A voice from way off in space called Always back from the bottomless well into which he had sunk. "Sir, it's three-twenty. It's time for you to get up." Sharp talking.

Always' tongue must have swollen three times its normal size. Either that or he had a sock stuffed in his mouth. Sand and dried sweat combined to seal his eyes shut. He was struggling to come out of his sleep. A chill shook his body in the relative cool of the desert predawn. He regretted he had not put a poncho over himself when he laid down. The stitches over his eye throbbed, and his head felt like it would split open as he brought himself to a sitting position. Altogether, he felt like he was on the losing end of a fifteen-round decision.

"Have you set the radios, Sharp?" Always would never have dropped the rank if he had been more awake. It was a rule he set for himself.

"Yes, sir."

"What's my call sign?"

"You're Romeo 36 today, sir. Here's the 'cheat sheet.' " Sharp handed him the board with markings of the battalion's call signs and frequencies.

Always tried to memorize the critical alphanumerics as he relieved himself, his eyes straining in the faint moonlight.

Wearily he climbed into the commander's cupola to establish radio contact. He was tempted to walk back over to the TOC for the exercise, but knew he had to ensure that he could talk to everyone from his own radios. Even the artillery officer and the air force liaison, themselves just getting up into the personnel carrier only fifteen feet away, would be compelled to check in over the radio.

While answering the colonel's question Sharp had stripped away the sleeping gear, cleared the jeep and the Bradley for action, and departed with the jeep. Everyone in the battalion was ready to move.

Two scout vehicles had been fired upon in the night, one apparently destroyed since no further word had come from it after the initial report of contact. The other had escaped after being chased back from the vicinity of Hill 781. It had been able to drop off a dismount team with a radio, smoke, and a panel marker to bring in the airmobile assault. Precise locations had been given on the enemy vehicles on Objective HOOK. Always told his artillery officer to plan a fire mission to put in on them at precisely 0415, the time he figured Team Bravo would be deploying for its assault. The winds were as predicted; Always confirmed the smoke missions. With a final response from all parties on the net, Always closed out the conversation and moved with Delta to an attack position just short of the line of departure. At 0400 the battalion started to move across.

The smoke covered the attackers well and blew to the southeast, hiding their advance up the valley. It had a slowing effect on the movements of the columns, but the ground, particularly along Axis LEFT (Always had made his reconnaissance along Axis RIGHT the day before), was so rough as to force them to a five-mile-per-hour pace. The combined effects were advantageous to the task force. It was moving unheard and unseen along its attack routes with no interference from the enemy. Captain

Baker, on his toes this morning, called and asked to defer the artillery strike until 0430. Always approved and passed the order to his FSO.

Remarkably, everyone's radio was operating in secure. The colonel was thankful for small miracles. He had been driven almost mad switching back and forth the day before. As bad as his head ached this morning, he did not want to repeat that grief. He munched on a hunk of chocolate cake from his MRE stock. His only difficulty at the moment was the beating he was taking in the cupola as the vehicle lurched up and down a maze of wadis wildly intersecting the valley floor. A belt held him in his collapsible seat, exacerbating Always' savage pivoting as it pressed against his hips. He didn't know how Sergeant Kelso could stand to keep his eye pressed against the gunner's sight under these conditions. The colonel concluded his gunner was one tough hombre.

Baker hit HOOK just as the artillery lifted and before Carter emerged from the smoke to come under the guns of the enemy. The ambushers were ambushed, and Baker did a good job in mopping them up. The three BMPs that survived the artillery concentration were destroyed by the approaching Bradleys, who picked them out with their thermal sights through the smoke. The defenders never saw what hit them. The only setback came when a dismounted Sagger missile hit one of Baker's Bradleys, destroying the vehicle, wounding the gunner, and killing three of the infantrymen riding in back. Instinctively, Always knew that it was a mistake to ride the infantrymen into close combat like that. The Bradley was a tough vehicle, but anything that moves can be destroyed if something big enough hits it. The tactics should have been a little more carefully worked out. There was no reason why they could not have dismounted a little earlier, at least before they emerged from the smoke. He recorded that thought in the back of his mind. There would be casualties enough in the battle. No need giving any away.

The airmobile assault had lifted off on time. A report came in that they had landed all right, but shortly after, that radio communication had been lost. The range was too great, and the infantrymen were staying much too low—as they must—to keep the radio waves on line. Always hoped Bravo could reestablish communication as he closed in on its infantrymen.

Brigade called at 0500 to report that the armor battalion on the right was being held up by a stiff defense by the enemy. The higher headquarters also reported a buildup of enemy forces to the north of Always' objectives. For the moment, the orders still stood, but the brigade commander implied that a change might be forthcoming. The situation would have to develop a bit before any reassessments could be made.

In the vicinity of Checkpoint 3 an obstacle was located that had not been found by the scouts during the night. Cautiously, Team Delta deployed its infantry dismounts into the high ground to the west. The conservatism paid off as they found an enemy scout after a forty-five-minute hunt, drove him to ground, and killed him. A tough bastard, he had fought to the last, in the final moments throwing his radio off a cliff in order to protect whatever frequency it had been set on. The crush of time negated sending down a team to retrieve it.

Subsequently, the engineers came up and removed the mines from in front of the obstacle, allowing the task force to resume its movement. An artillery barrage came in, but missed by a wide margin; the dead enemy scout was to have adjusted it. Although the attack was being slowed (the supporting attack was held up to allow for coordinated movement), there was no way around the mine field. Every pass through the wadis at this juncture had been similarly mined. Always would have to reduce them methodically.

Nonetheless, things were going fairly smoothly. With only a few casualties, the task force had moved approximately half the distance planned. The ambush had been nullified and the obstacles were being reduced. Only the lack of communication with the infantry on CHOPPER bothered Always. He was reassured, however, by the fact that the TOC reported that all helicopters had made it back safely. They had gone in unharmed. Yet, time was slipping away. It was almost 0700, and they still had a way to go. For mechanized forces that were able to sprint at more than forty miles an hour, they were moving at a snail's pace. Always tried to control his impatience.

He did not know, could not know, how critical time was to his mission. The armor battalion on his right was suffering heavy casualties at the moment, faltering in its attack, and relieving enemy concerns from that direction. Having deduced that Always was making progress, albeit slow, the enemy shifted forces to defeat him before he could mount his final assault. The defenders had seen the airmobile insertion, and although they could not react immediately, they were mustering two motorized platoons with an attached section of tanks to destroy Bravo's infantry. In the meantime they made life as uncomfortable as they could for the dismounted force by plastering the area with mortar and artillery fire. Lieutenant Rodriguez, in command of the two-platoon force, was having to move out of the fire, strapped with increasing numbers of casualties. His attempt to reach his commander on the radio met with repeated failures. It was only by dint of his strong leadership that he held his men together, determined to press on with his mission.

At 0800 the attack helicopters appeared, approaching from the task force's left rear.

"Romeo 36, this is Sierra 82. I've got your dust in sight. What is your situation?" It was the lead company commander.

"Shit!" Always cursed over the intercom. The aviator was talking on his radio in the unsecure mode (in the "red").

"This is Romeo 36. Go secure. I repeat, go green." Always had switched his radio off secure.

"Uh, negative. Can't do that. We don't have your cipher. I've got to stay red. Where do you want us?"

Always was furious. Trying to control his temper, he told the captain he was not ready for his fires yet, that he estimated it would be another forty-five minutes before he was in position to assault, and that he needed the aviators to get the right cipher before coming up on the battalion net.

The aviator responded that he could not burn fuel for the next forty-five minutes and still have enough for the fight, that he was not sure he could find the cipher, and that he would have to go back and set down to conserve fuel. With that he led his element back down the valley.

The whole conversation had taken only two minutes. But it was enough. The enemy picked it up, and although the eavesdroppers could not hear the subsequent conversations when Always went back to secure, they had the frequency. They would know whenever he was talking by the break in squelch. They now had some options. They could allow the broadcasts to continue, thereby picking up vital information should Always' battalion return to the unsecure mode (as they had the previous morning); jam them off the net at the critical moment; or, if they could produce an operator with the voice inflections they had picked up from Lieutenant Rodriguez earlier, they could try some imitative deception. They had learned from their sources around Purgatory that Lieutenant Colonel Always had a soft spot for his soldiers. Perhaps they could play on that compassion to induce him to blunder into a desperate attempt to save Rodriguez.

"Romeo 38, this is Alpha 38." It was the brigade commander calling.

"This is Romeo 38, over."

"Roger. Look, we got big trouble developing. I'm calling off the attack on 781. Our whole line is folding back on the right. I'm shifting the armor down into the valley you just came from. I want you to hold up over on the east by Checkpoint 4. Expect an attack from a reinforced motorized rifle battalion coming out of the north. We've got a report that several battalions are entering the picture rapidly."

Always acknowledged the transmission, received a stern warning to hold his position at CP 4, and began to order his units to shift. But it is always difficult to shift a mechanized force in midstride. As Always signaled Charlie and Bravo to move to CP 4, the enemy decided to make his life a bit more complex. He jammed Always off the air.

The static buzzed through his eardrums like a saw. Every time he spoke he received a head-wrenching screech. The receiving stations were trying to answer, getting out a maddening few words, only to have the essence of the message drowned out. Always tried to fight through it. He had to get everybody moving quickly. This was no time to be shifting frequencies. At that moment, Lieutenant Rodriguez came up on the battalion net, transmitting in the red.

"Romeo 36, this is November 25, over." Always heard the call clearly, never suspecting that Rodriguez was being helped by the enemy who was retransmitting the lieutenant's call for help, ensuring it would get through to the colonel. It was a minute before Always identified the platoon leader's call sign from his cheat sheet.

"This is Romeo 36. Send your message."

"This is November 25. I need to know when you're getting up to me. I'm being chopped up by enemy artillery and I can hear some tracks approaching from the north, over."

Always felt caught in a vise. He was sorely tempted to divert Team Bravo or Charlie to pick up the infantrymen. He had met Rodriguez, and he knew he would not be calling for help unless he were in trouble. But his brigade commander had expressed an urgency to the mission to defend in the vicinity of CP 4. He responded, "How many antitank weapons do you have?"

"This is November 25. I have six Dragons and six LAWs, over."

The enemy monitored the report and sent the information on to his own battalion. Always made a quick assessment. The LAWs (light antitank weapons) would be of little to no use. The Dragons were a possibility, but they were unreliable. Six of them would not go very far. But there was no alternative. "Get yourself to ground. There will be no help reaching you for some time. I'll get to you as soon as I can. In the meantime, try to get lost to the enemy. If you can break contact and get out, we will be holding in the vicinity of Checkpoint 4."

It took the enemy three minutes to deduce where Checkpoint 4 was located.

Rodriguez was stunned by the news. He looked around for a piece of ground where enemy armor could not approach, found nothing that suited him, picked the next best spot, and ordered his men to move to it. He was burdened with five casualties now, three of whom could not move without help. He was troubled by the one KIA (killed in action)—he had learned never to abandon his dead. But he had only thirty-three men left, counting himself and the two medics. He hesitated, then called to his platoon sergeant. "Mark Jones's body. We'll come back for him when we can."

The battalion radio net was a madhouse now, a whirling, screeching, buzzing, ringing cacophony. Always gave the code word to move to the alternate frequency, "Bayonet!"

The TOC barely heard it over the jamming and began repeating it over the radio. By its standard procedures the TOC would leave one radio on the old net for a few minutes while shifting the others to the new frequency. Always was concerned about the infantrymen on CHOPPER and ordered the TOC to get a radio up in the red (in the unsecure mode) to continue listening for them. For ten minutes confusion raged over the nets as the subordinate elements scrambled around trying to figure out if and when they should shift frequencies. Those few minutes were fatal.

While the madness on the radio nets was unfolding, the air force lieutenant came scrambling up atop Always' Bradley. "What do you want, Lieutenant Smith?"

"Sir, I couldn't reach you on the radio. I got a spot report from some of our fast movers (fixed-wing aircraft) that enemy vehicles are moving down the pass from the northwest toward CP 4."

"Are you sure?" That was the flank Always was concerned about, but it did not seem likely that the enemy would approach from that direction. The brigade commander had indicated the enemy was coming from the north. It was possible, however, that he had meant the northeast, or that his intelligence was faulty.

"Yes, sir, I'm sure. I asked them to confirm and they did. They estimated about one battalion."

"Okay. Get back to your track and tell the artillery officer to put in the scatterable mine field at the mouth of the pass."

On the alternate frequency the scout was reporting enemy movement from the north, with a large dust column coming out of the pass at 781. Simultaneously, the battalion executive officer called in a report that the artillery was pulling back into the valley they had left that morning. The artillery pieces had jammed up the routes, interrupting the following battalion trains. Ominously, the artillerymen were reporting enemy approaching from the southeast. Major Walters needed to know where Always wanted to move the trains, and if he wanted the TOC to move to CP 4.

Always held the trains and the TOC back in the vicinity of the LD. The battle was in too much a state of flux to risk piling them into CP 4. With all the conflicting reports he had received he did not know from which direction he was about to be attacked. Brigade was currently being jammed off the air. Always would have to figure it out for himself.

E Company was the first to reach CP 4. Evans had gotten only a part of the message to move there on the original frequency. After that he had heard nothing, not the call to change frequency or the call over the artillery net that the mine field was going in. The message was never passed over the battalion net. Two of Evans' lead vehicles exploded as they rushed into CP 4. The mines worked.

Lieutenant Colonel Always led Alpha and Delta into an arch north of CP 4, oriented to the north and northeast. After raising Captain Evans on his internal net and learning of the disaster in the mine field, he had them face his two remaining antitank vehicles at the exit from the pass. He then brought in Bravo and Delta, facing them to the east and southeast, respectively. As the battalion commander literally circled his wagons, the first wave of the enemy appeared, coming at him from 3,500 meters out of the north.

The helicopters chose this moment to return. They had indeed discovered the correct cipher, but were not aware that the task force they were supporting had shifted frequency. Although the TOC had stayed up on the old frequency, it was operating in the red for the sake of Lieutenant Rodriguez. The aviators, not knowing this and talking in the green, were completely unheard. Nor did they have the alternate frequency.

Their captain, being aggressive, understanding the original mission, and seeing all the dust being kicked up north of CP 4, assumed that the attack on JAB was being kicked off. He brought his company in to cover the attack, thereby bringing his helicopters directly over the attacking enemy, assuming it was Always' people below him. From this confusion stemmed the slaughter that followed. As the aviators realized they were directly above the enemy they opened up in desperation. In two minutes fourteen enemy vehicles had been destroyed—six tanks; seven BMPs; and a ZSU-34, an antiaircraft gun track. So too were twelve attack helicopters destroyed, along with two scout helicopters. The ranges had been so short that the aviators never had a chance to evade the fire.

Always did not know the reason for the melee he was observing to his front. He was both admiring of the courage of the pilots and aghast at their recklessness. With their lives they had bought him a desperate few minutes to set up his defenses. It would be a direct fire exchange on the ground now. The artillery was on the move, pulling back from the pressure on the west, and was not yet deployed in a position from which to support. Mortars would be of little help in this armored exchange. Tank and Bradley gunnery would count for everything.

At that instant the chemical strike covered them. Always was leaning outside his cupola yelling at Captain Archer to pick up the enemy at 3,000 meters when the artillery-borne gas exploded all around him. He struggled to pull his body back inside the hatch, choosing to slam the cover before pulling on his mask. When he yelled "gas" to his crew a small whiff entered his throat, searing his lungs and convulsing his stomach. As he pulled the mask tightly over his face, the vomit gushed into the rubber-enclosed space in front of his mouth, filling the mask to the eye sockets. His body twitched, his eyes watered, and he vomited again. He fought to clear his mask and get some air into his lungs. A third wretch and his mask was almost full with his own vomit. As he took his first breath he flooded his nostrils with his own puke. He broke the seal at the bottom of his chin and let the slop spill out. With great effort he brought his body back under control.

"Are you all right, sir?" Sergeant Kelso asked with a great deal of concern in his voice.

"Yeah, I'm okay. Check Spivey." Always was trying to

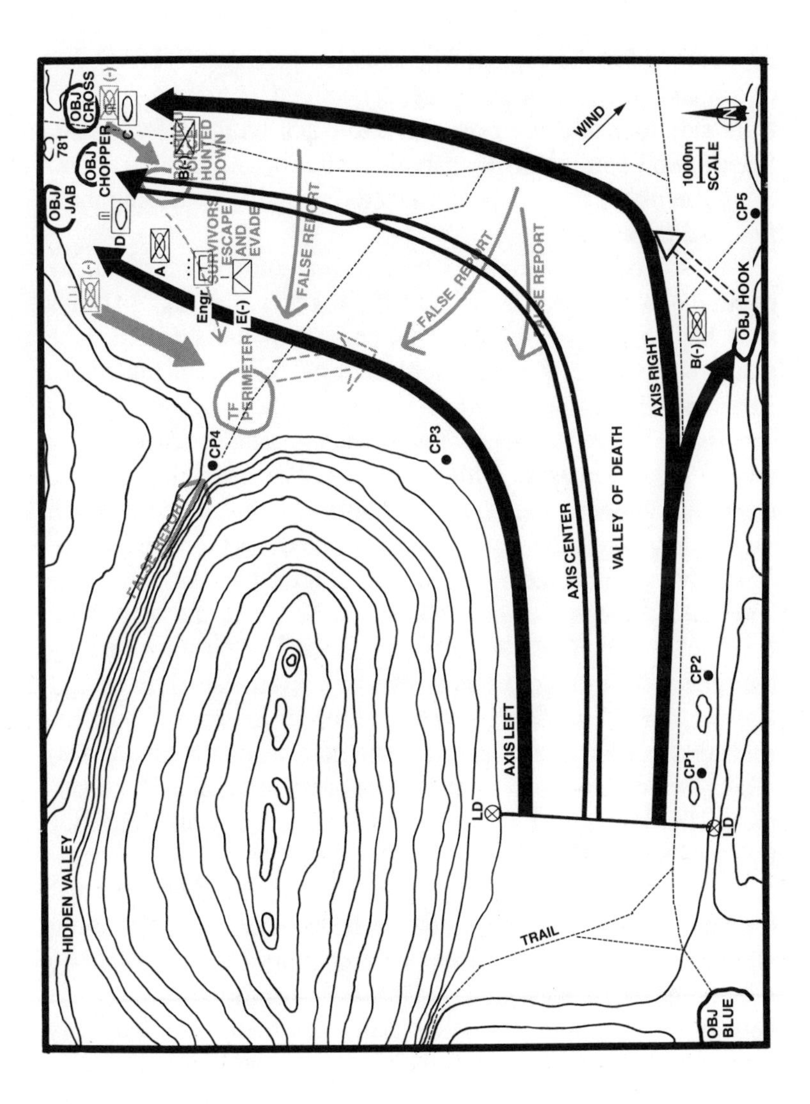
OBJ CROSS
781
OBJ CHOPPER
OBJ JAB
C
D
A
B(-)
HUNTED DOWN
SURVIVORS ESCAPE AND EVADE
Engr
E(-)
FALSE REPORT
FALSE REPORT
FALSE REPORT
FALSE REPORT
TF PERIMETER
CP4
CP3
CP1
CP2
CP5
WIND
1000m
SCALE
OBJ HOOK
B(-)
AXIS RIGHT
VALLEY OF DEATH
AXIS CENTER
AXIS LEFT
LD
LD
TRAIL
HIDDEN VALLEY
OBJ BLUE

move around in the cupola. The cramped space was inhospitable to a man encased in a gas mask. The slightest turn of the head from side to side caused the mask to catch on the sides, radio, latches, and corners of the paraphernalia located in the cupola.

"I'm okay, sir," Spivey answered, his voice sounding muffled and far away.

"Get ready for action. There's a battalion's worth of enemy coming down on us. They'll be here in a few minutes." Always now had to prepare for personal combat at the same time he struggled to focus his battalion's efforts on the enemy force. Reaching all of the subordinate units on the radio was nearly impossible. Many of them were not understandable when talking out of the gas masks. Although the CVCs had a hookup for the mask, allowing the voice to be directly transmitted over the radio, they often did not work or worked only on one radio. Since every commander was up on at least two nets, one was addressed through the microphone on the CVC, the other on a hand-held mike. The latter was audible only when stuck directly under the chin or onto the throat, and even then just barely. Always found himself yelling at the top of his lungs to get his messages across. If he received simultaneous calls on the two radios, which happened often, he was certain to understand neither. The donning of the gas mask had reduced his efficiency by more than half.

With only two of his companies facing the oncoming enemy, the others disposed to face attacks from other directions (after all, reports had indicated they were coming from all points on the compass), it was going to be a desperate shoot-out. The helicopters had helped a great deal with their kills, but they were out of the picture now, their survivors pulling out, unable to establish communications, unsure of operations on the ground, and not inclined to risk a repeat of the grave losses they had already suffered. As the Bradleys put up their TOW missile launchers, the tanks ranged the oncoming enemy. At 2,500 meters the shooting began.

The T-72s and BMPs were appearing for only a few seconds at a time, rising over the small ridges running between the lines of wadis perpendicular to their front and disappearing as they descended into the next wadi. This negated the effects of the TOWs, which needed an unbroken line of sight on the enemy for the duration of the missile's flight, more than ten seconds at 2,500 meters. For the tanks it was another story. As soon as they had the target in their cross hairs they could kill in a split second. The same was true for the 25mm gun on the Bradley, although it could not hope to stop a tank. Always recognized this and tried to direct the fires accordingly, tank against tank, Bradley against BMP.

The enemy's speed was amazing. Within minutes he was at 2,000 meters and closing. Some of his number had been reduced, but at decreasing ranges the advantages were passing to the enemy. The defender's gunnery was severely reduced by the effects of the chemical strike. It was impossible to get the shooter's eye close enough to the eyepiece for a clear, sustained picture of the attacker. Moreover, the dust the movement and firing were kicking up further exacerbated the conditions.

Always saw a BMP to his right front at 1,200 meters. He called to Kelso and shifted the gun in the general direction. "Got it," the gunner called. He received the order to fire from Always and immediately opened up. The first round kicked into the dust short and to the left. The next four went through its armor forward of the midpoint of the vehicle. The BMP took a crazy veer to the right, crashed into an embankment, and came to a halt.

In the next instant a T-72 crashed a round at the battalion commander's Bradley, barely missing it, the projectile passing by high and to the left by a few inches.

Always yelled to Spivey, "Back off! Get some smoke out."

The driver reacted fast. In a flash he peeled back down into a nearby wadi, covering his move with smoke from diesel fuel splashed across the hot engine exhaust.

For a moment Always was disoriented. He ordered Spivey to poke the nose of the Bradley cautiously back up over the rise. Inch by inch the driver pulled the big vehicle up until the commander ordered it to halt.

The enemy was now closing to 1,000 meters and still coming on. Always was ignoring calls from all parties but Alpha and Delta. He estimated that they had each lost more than a third of their vehicles. He sent a warning order to Captain Carter to be prepared to orient to the north and pick up the incoming enemy.

Suddenly the enemy's close air support came in and rocked the ground around Checkpoint 4. One of Echo Company's antitank vehicles went up in flames, along with the communications vehicle behind the company XO. Now only one antitank system was covering the pass. Bravo Company had lost two vehicles and had a third damaged. Charlie lost a tank. The antiaircraft guns opened up, the gunners hampered by their chemical gear, filling the air with lead nonetheless. Bravo Company took a chance, lowered a ramp, and unhorsed its Stinger team. The air defense men struggled to pull their missile out of the crowded vehicle and get it into action, but by the time they had reacted the aircraft had pulled away. Given the lethality of the area they were in, its chemical contamination, and the movement of friendly vehicles every which way, they abandoned their effort and returned to the track.

The second enemy aircraft run, three minutes later, was more even. A Bradley and a tank from Alpha Company exploded, but two aircraft were hit, one crashing into the mountain behind the action, the second pulling away with a dark cloud of smoke trailing behind.

Always looked to the north. The first enemy battalion was now within 750 meters and coming on fast. Behind it in the distance came the second battalion. The colonel called to Bravo and Charlie to add their fires against the approaching enemy. At that instant the enemy spewed out columns of smoke from the lead vehicles, broke to the west, and swung wide behind the battalion, passing on to the south. A few more shots were exchanged, then he was gone. In the wake came the second battalion, attempting to conceal himself in the smoke of his brother element. Simultaneously, the enemy artillery intensified its fire. No more chemical rounds were being fired now; it was all high explosive.

The din in and around Checkpoint 4 was unnerving. Incoming and outgoing ordnance was raising dust everywhere. Burning diesel blackened the sky. Wounded men were crawling away from wrecked vehicles, seeking shelter form the artillery and the charging tanks and BMPs. Medics struggled in their gas masks to get help to the wounded. The artillery filled the air with shrieking whistles, punctuated by deafening explosions, competing with the clamor of the 105mm tank guns; the whooshing antitank missiles; and the heavy, machine-gun-like staccato of the 25mm Bradley guns.

In fifteen minutes it was over. The two attacking battalions had passed through to the south, approximately one company surviving from the first, slightly more than two from the second. They were now falling upon Always' brigade, smashing into the retreating armor battalion and chasing down the rear of the direct support artillery battalion.

Lieutenant Colonel Always called brigade headquarters and reported that he had failed to stop the attacking enemy, although he still held Checkpoint 4. His commander, preoccupied with the fight that was breaking elsewhere in his sector, ordered Always to hold where he was and ensure that no further elements got by him. The last words were spoken abruptly, and Always felt their sting. For the second day in a row he had failed to accomplish the essence of his mission. By letting the enemy companies slip by he had jeopardized the entire brigade. He set himself to preparing for yet another defense, not certain from which direction the enemy was liable to come. Any thought he had of sending out an element to try to save Rodriguez was stifled. For the next three hours he redistributed ammunition, evacuated his wounded, and improved defenses. He waited for an enemy that never came.

Lieutenant Rodriguez was having a bad day. The dash for defensible ground had met with only partial success. The casualties held up the movement, and eventually a trail party of five able-bodied men went to ground, doing their best to protect their wounded comrades. The lieutenant herded the remainder of his two platoons into a canyon and had them start digging in. For fifteen minutes he kept radio contact with his trail element; after that there was some firing and they went off the air.

By this time Rodriguez's men had run out of water, their exertions giving rise to a maddening thirst. Several pints of sweat were oozing out of each man every hour, and as the sun rose in the sky it baked one and all unmercifully. The lieutenant set up his antitank weapons in pairs, layering them in depth up the mouth of the canyon. It was an hour before the first enemy tank came nosing in. One Dragon careened off erratically; the second found the enemy armor and blew into it, killing the commander and gunner. The enemy withdrew in a hurry, but only for a moment. As he deployed his infantry to work around the top of the canyon, he began to pepper Rodriguez and his men with mortars. In another hour and a half, small arms fire had pinned down the Americans. The mortar fire now was being accurately adjusted onto the small pockets of men waiting in defense.

Rodriguez turned to Sergeant First Class Peterson, his platoon sergeant. "They'll be coming back in here again pretty soon. We'll give it a try, but I don't think we can hold. If it gets untenable, take whomever you can and try to break out for CP 4."

There had been no contact with battalion on the radio for hours. Neither platoon leader nor platoon sergeant was certain the battalion even existed anymore, much less if it were waiting for them at CP 4. Trying to break out to the battalion, therefore, was a long shot, but it was probably the only chance. The infantrymen would have to exfiltrate by ones and twos. Any group bigger than that would be hunted down and slaughtered like so many jackrabbits. A man on foot in the desert was easy prey.

The tempo of small arms fire and artillery picked up as the tanks began inching their way back in. Two LAWs were fired with no effect other than to bring the tanks' machine guns in on the infantrymen. The latter were torn apart instantly. A Dragon gunner moved to the awkward but prescribed sitting firing position. A mortar round caught him in the open before he could release, puncturing the missile container and rendering it useless. A second Dragon missed. A third and fourth hit a BMP and a tank, destroying the former and immobilizing the latter. The last four LAWs did no damage. With the infantrymen now out of antitank ammunition it was a duck shoot. With increasing boldness the enemy tracks rolled up and eliminated each pocket in turn. Rodriguez yelled the order to break contact and get out, then began firing his M16 wildly at the lead tank.

A machine gun round broke his arm at the elbow as he dropped his three remaining smoke grenades in an effort to cover his retreating men. Bleeding heavily he scooted to higher ground farther up the canyon. A second round chipped out his hip joint and careened through his buttocks. He crawled through the dust, coming to rest behind some rocks. He lay here for forty-five minutes, unfound by the enemy. At first his thoughts were preoccupied with the pain in his hip and an overwhelming need for water, but later they drifted to serene memories of his childhood. He died shortly after he was taken prisoner.

Two hours later Sergeant First Class Peterson made it to Checkpoint 4 with two other infantrymen. He was brought to Lieutenant Colonel Always, where he gave his report. The colonel took it stoically, stifling his anguish at the loss of so many men to no end. He realized how foolish it had been to put them so far out beyond support. Later that afternoon four more men straggled in. That was all.

At about 1400, orders came for a night withdrawal. The brigade had been rolled back on the flank, and intelligence indicated that the enemy was continuing to build his strength to the northeast. The task force would clear the mine field in the pass and move some twenty-five kilometers to the northeast. The report of enemy in the pass was false. In fact no one would admit having made such a report, much to the chagrin of the air force liaison officer.

By midafternoon the battalion's leaders were assembled for an after-action review. Lieutenant Colonel Drivon and his team were extremely thorough. The aviation commander, who joined the assembled meeting, turned a little red at some of the comments. So did the artillery commander. Always was slightly relieved to have company on the receiving end of the faultfinding, but he knew once again that the failures were mostly his. Later that night as he led his battalion through the dark into its new defensive position, he had a chance to review the lessons of the day:

Integrate all supporting elements with the actions of the ground maneuver force. Aviation must understand that orders are liable to change and that flexibility is a prerequisite for success. Artillery must never fail to be in position to

support the maneuver element, whether it be attacking, defending, or conducting retrograde operations.

All elements must remain in communication. Supporting units must shoulder the burden of keeping in constant touch with the supported element. When orders change and frequencies are switched, a thorough dissemination of the changes must be made. The maneuver commander must insist vigorously on an ironclad system of keeping abreast of a changing situation.

Never put infantrymen on the floor of the desert unsupported in a pitched fight against armored forces. They cannot sustain themselves for long; very quickly the balance will shift to the heavier force. Set up communications relays to keep them in contact with their parent headquarters. Toughness cannot compensate for poor planning and foolish decisions.

In the final analysis, gunnery is the decisive factor in armor battles. Fire control and marksmanship are twin pillars of victory in the close engagement. Practice gunnery under battlefield conditions; make it realistic by arranging for burning hulks, widespread smoke, dust, movement, and loud noises. Get used to firing while wearing gas masks. Bore sight main guns at every opportunity. Designate sectors and targets for each weapons system. Fire fast, and make every shot count.

Early warning of enemy air attack is imperative. It does no good to react to an air strike after it has already hit. By that time the enemy has dropped his ordnance and is gone.

Be ready for a chemical attack at any time. Be able to function under the most extreme conditions, but know when to come out of gas masks. The most well-trained units will suffer inefficiencies when encased in chemical gear. Don't

stay in masks a minute more than is necessary, particularly when in a close engagement. But don't uncover prematurely either.

Don't believe every report passed to you in the midst of battle. Weigh every report against its plausibility. When in doubt orient on the enemy, with one wary eye for the unexpected. Concentrate firepower at the decisive point.

These and other thoughts were running through Always' mind when he arrived at the release point and had to divert his attention to moving the task force into its defensive sector during the few hours before dawn.

CHAPTER 4

Defense in Sector

The battalion's defensive sector was enormous. The front stretched almost 8,000 meters from northeast to southwest, tying in to mountain ranges on either flank but offering the enemy ample ground over which to approach. The center of the sector was a flat stretch of desert, split by one gigantic wadi that cut from the southwest toward the key terrain in the northern half of the battle area, Hill 910. Always' orders established a no-pass line deep in the rear of his sector, almost thirteen kilometers behind the front. He could expect to be attacked by a motorized rifle regiment, probably within thirty hours of his occupation.

It was a difficult tactical problem. As open as the desert seemed, there were several options available to the enemy. The five combat companies at Always' disposal seemed inadequate to cover the frontage and depth assigned to his unit. Deep in his rear, but still in front of the no-pass line, designated Phase Line STOP by Always, the ground broke up into countless deep ravines. Any one of them offered a major route through the depth of the sector. It would be impossible to cover all of them.

Try as he might to spread his companies over the expanse, he could not adequately cover all of the ground and have any depth. Therefore, he decided to try an unconventional approach, creating a sixth company by dividing up his existing units, reducing their number of organic platoons, and putting the remainder under the command of Captain Coving, the Headquarters Company commander. He was the most experienced commander in the battalion, having previously commanded a line company. He gave him the S-3's Bradley for his command vehicle. Major Rogers could operate from the TOC, since this was a defense in sector and movement would be limited.

The defensive area was divided up into three belts. The initial defense was at Phase Line FORWARD, covered in the north by Captain Evans with two antitank platoons (his third platoon had returned from the armored battalion and was now attached to Captain Coving) and in the south by Captain Baker with one tank and one Bradley platoon. The second belt at Phase Line MIDFIELD consisted of Captain Carter in the center and south with one Bradley and two tank platoons, and Captain Archer in the north with one tank and two Bradley platoons. The final belt was in front of STOP, Team Coving in the north with one antitank and one Bradley platoon, and Captain Dilger in the south with one Bradley and two tank platoons.

This much Always had been able to decide from a map reconnaissance, but he was experienced enough to know that all this did was get the companies into general positions with the requisite composition of platoons. It was still dark as the units were moving in, and he would have to order an immediate review at first light so that he could flesh out this simple plan in time to accomplish the immense work load that a defense entails. At that time he would have only twenty-four hours remaining to put in a defense that would have to stop a force three times its size, a force reinforced by artillery that vastly outnumbered his own, and with ample options as to the point of attack. Always knew he would have to develop a defense in depth, and that the only way this could be accomplished with his paucity of resources would be by developing a plan of movement that allowed the companies to leapfrog backward as the battle unfolded. His appreciation of the complexity of combined arms warfare had grown greatly in the past several days. Now the ground and the anticipated enemy size would put him to his severest test.

Major Walters had done a tremendous job of moving up the broken and disabled vehicles leftover from the last battle. Within minutes of establishment of the unit trains, work commenced on repairing and returning combat vehicles to the line. Simultaneously, the refueling operations ensured that each unit moved into its position with full fuel tanks. A less professional battalion would have waited until dawn to achieve these difficult operations, thereby stealing critical time from the company commanders who needed to be establishing work priorities in their companies. Again Major Walters quietly and efficiently gave the task force the opportunity to focus on the tactical matters at hand by efficiently integrating the details of combat service support.

While Command Sergeant Major Hope retraced the route of march, policing up the pockets of soldiers and vehicles that had become misoriented in the dark and reporting to the maintenance trail party the location of equipment that had broken down on the move, Major Rogers set up the tactical operations center to begin immediate planning for the sector defense. It was to this location that Lieutenant Colonel Always pulled in at 0300 to meet with his staff. Under the lightproof expanse of canvas, the maps were unfolded and the staff assembled to give their initial estimate of the situation.

Always was now fighting fatigue. He had poured cup after cup of black coffee down his throat until his mouth was heavy with the bitter taste, yet he still found it hard to focus. His staff looked as rocky as he felt, and he knew that the exhaustion he was experiencing had permeated throughout the command. The observers had made a point of counseling him on the need for a "sleep plan"; yet he knew that the time to sleep was not at hand. Modern combat had created a dilemma for the commander that was not to be solved by doctrine on sleep plans. Equipment that could function and allow operations around the clock was operated by men who still had the same biological needs as their prehistoric ancestors. Always reached down into himself for reserves and tried to focus on the matters at hand. He hoped that he could find the energy not only to sustain himself but to pass on to the men around him.

Defense is inherently the stronger form of war (or so Clausewitz said). The defender has the advantage of knowing the terrain better than the enemy (after all, he occupies it), of digging in and selecting the terrain from which he will defend and on which he will bring his weapons systems to bear. It gives him time to alter ground to suit his purposes, to lay in mine fields and obstacles, to register his artillery, and to sight his weapons. Done correctly, the defense will force the attacker to pay dearly for the right of passage over the defended terrain.

But the attacker is not without opportunities. He can choose the point and time of his attack, overcoming the defender who might unwisely stretch his defense (as the saying goes, ". . . he who would defend everywhere is strong nowhere"). He can concentrate his forces so that even a resolute defense can be overcome at a vital point, unhinging the defensive plan as the attacker erupts through a penetration. His ability to do so can be enhanced by good intelligence as to the dispositions of the defender, intelligence that can best be achieved through an aggressive and thorough reconnaissance effort. This concern was driven home to Always by his intelligence officer, who voiced his suspicions that, even as they spoke, enemy observation posts were set in hiding overlooking their positions. He made a strong appeal for the task force to commit substantial resources to the counterreconnaissance effort.

As in all things, however, such a decision was not easily

made, for the engineer briefed just as convincingly a need for the obstacle work to be performed by the maneuver elements. He had a good point in that his own thin forces could not accomplish, without help, the extensive work that would have to be done. A single platoon could hardly move to the many positions to be defended, prepare tank ditches, dig in multiton machines, improve routes for the mobile phase of the defense, lay and record mine fields, and stretch wire over the miles and miles of ground from which the enemy could pick and choose for his point of concentration. Any forces committed to the reconnaissance battle would be taken away from the engineer effort, which would ultimately determine the viability of the battalion's defense.

The companies also had a need for their men. Ammunition would be coming up shortly after dawn, necessitating distribution and reloading. Extra stocks would have to be dug in with overhead cover and situated where they could be reached easily in a fight, which meant that not only the initial fighting positions would have to be prepared but supplemental ones as well. Infantrymen would have to dig in overhead cover, sight their weapons, walk their sectors, and rehearse their movements. Equipment would have to be maintained and weapons stripped and cleaned, rezeroed, and bore sighted. Patrols were necessary to keep the enemy from getting a foothold in the defensive lines, particularly under cover of darkness. Obstacles would not only have to be put in, they would have to be guarded lest the enemy undo them. Time and time again it had been shown that an uncovered obstacle was no obstacle at all. Vehicles would have to reconnoiter their routes of movement from primary to alternate positions, and rehearse their maneuver during hours of light and darkness. Leaders would be involved in the preparation of orders and the reconnaissance and supervision that they involved. In short, there were more jobs than could be accomplished in the short amount of time allotted. Always would have to designate priorities and then entrust them to proper supervision by his subordinate commanders.

Through the predawn hours the staff members offered their views. The task force commander heard them through, gave his initial guidance, then departed for a first-light visit to his elements and a drive over the terrain he would have to defend.

The intelligence officer had been correct. Six enemy patrols had set themselves in place more than twenty-four hours before Always had arrived, selecting high ground that would be virtually impossible to reach by anything but the most determined foot patrols, but which afforded excellent observation throughout the depth of the sector. Their positions also afforded excellent communications back to the motorized rifle regiment already under receipt of its orders to conduct an attack. Reinforced with mobile reconnaissance teams from its division, the regiment had developed a plan to infiltrate scouts the next night into the defenders' positions, giving final confirmation of the defensive plan of its enemy. Armed with such knowledge, the regimental commander, a longtime resident of Purgatory, could modify his plan at the last minute to exploit any obvious weaknesses in the defense. He rested the majority of his men while he waited for the intelligence to be developed.

"We gave them a pretty good fight yesterday, didn't we, sir?" Specialist Sharp was as exuberant as ever.

"Yeah, we sure did." Always answered distractedly, his mind on the terrain he was studying.

Always envied Sharp his youth. The man never seemed to get tired, while he himself felt like he had been drugged. "Pull on over here for a minute. I need to shave before it gets much lighter." He resolved to look alert, even if he felt like death itself.

Sharp pulled the jeep to a stop beside a small hillock of lava rock and sand, immediately offering his commander his canteen for shaving water. He admired the older man's resilience.

He had felt that surely he himself would fall asleep at the wheel during the previous night's movement. It had been a struggle to stay awake at every stop along the route. Dawn had been a reprieve from his stupor, but he knew that shortly the sun would be baking into him, lulling him back toward his terrible craving for sleep. He wished he could be as alert as his battalion commander.

Unbeknownst to one another, each man quietly resolved to follow the example of the other, forcing a concentration on the mission at hand, stiffening their personal struggle with the elements pressing in on their biological needs for rest and recuperation.

Always learned much from his visits to the units. The company commanders learned from him his intentions for the defense while he learned from them how they would contribute to those intentions. Units were readjusted to take advantage of the terrain now revealed in the brilliant morning sun. Each made a case for access to the limited resources available for the mission. Baker got Always to commit the mortars directly to his company for the opening phase of the battle. Archer requested and received first priority on the bulldozers, already at work under the supervision of the bulldozer section sergeant. Evans was able to pass much of his concertina wire to Team Alpha, since the terrain Echo was overlooking did not offer much of an opportunity to lay in effective wire obstacles.

Carter extended his position farther to the south, tying in a platoon on Hill 899. Each company commander argued against committing any part of his unit to the effort of finding the suspected enemy outposts (OPs) in the surrounding mountains. There was too much work to be done in the immediate area of the defense. In his empathy for their concerns, a compassion perhaps intensified by his own fatigue, Always compromised his demands, limiting the counterreconnaissance requirement to a single platoon of infantry effort per maneuver company/team in the immediate

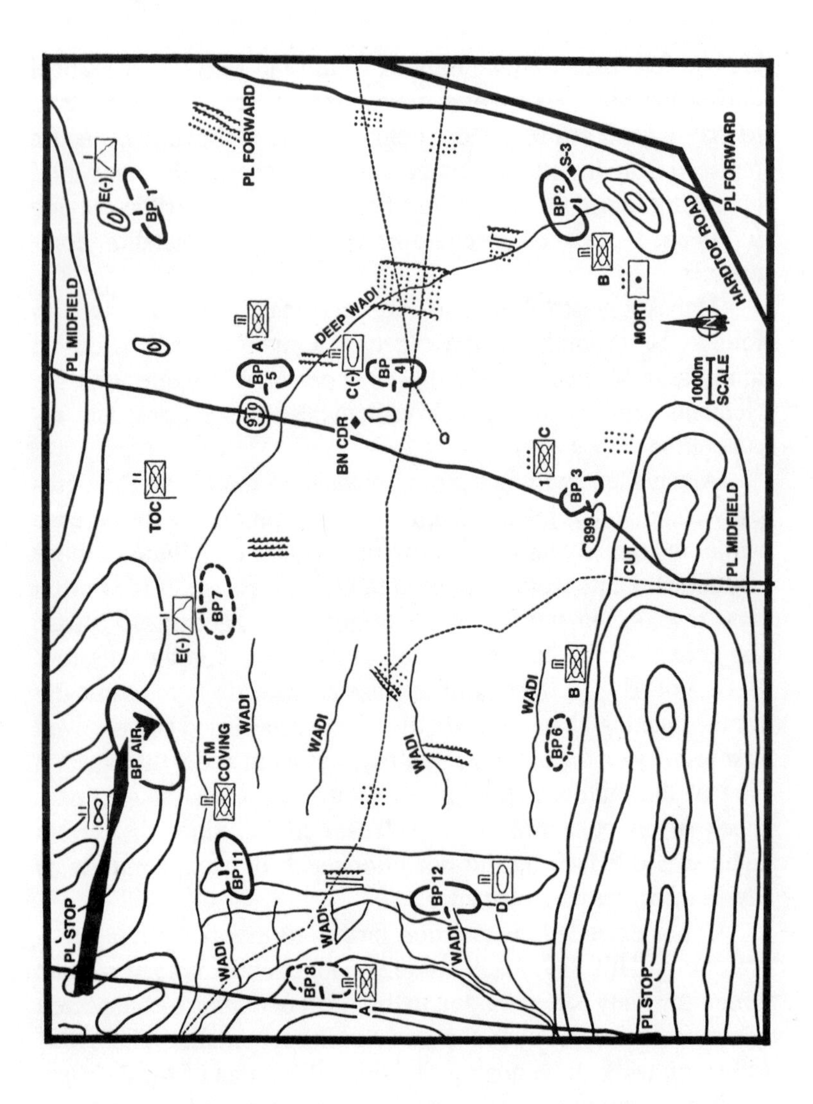
PL FORWARD
PL MIDFIELD
PL STOP
HARDTOP ROAD
DEEP WADI
WADI
BP 1
BP 2
BP 3
BP 4
BP 5
BP 6
BP 7
BP 8
BP 11
BP 12
BP AIR
E(-)
S-3
MORT
BN CDR
C(-)
TOC
TM COVING
CUT
899
910
1000m SCALE
A
B
C
D

Map 3. Sector Defense

Motorized rifle regiment takes southern avenue of approach after feinting in north and fixing Team B forward at BP 2.

Company E: Defend from BP 1; then BP 7
2 ITV platoons
Team Bravo: Defend from BP 2; then BP 6
1 Bradley platoon with infantry dismounts
1 tank platoon
mortar platoon (initially)
Team Charlie: Defend from BP 4—2 tank platoons
Defend from BP 3—1 Bradley platoon with infantry
Mortar platoon (when chopped)
Team Delta: Defend from BP 12
2 tank platoons
1 Bradley platoon with infantry dismounts
Team Coving: Defend from BP 11
1 Bradley platoon with dismounts
1 ITV platoon
Team Alpha: Defend from BP 5; then BP 8
2 Bradley platoons with infantry
1 tank platoon

Under battalion control: Scout Platoon, Engineer Platoon, Air Defense Platoon. Aviation gunship battalion in support.

vicinity of each battle position. This decision greatly perturbed the S-2 when he learned of it later in the morning.

At 1000 the orders group met at the TOC for ninety minutes while the plan was briefed and discussed. Always had found his second wind by then, and did a creditable job of inspiring his men for a vigorous defense. The essence of the plan lay in the details. Every subordinate would have to understand his role. More importantly, every man would have to execute it properly.

The three belts of the defense had been thickened by an intricate obstacle plan supplemented by a battle of movement. Always would bank on the superior speed of his tanks and Bradleys to outrace the enemy through the depth of his sector. Initially Echo and Bravo would decimate the enemy in the open ground behind Phase Line (PL) FORWARD. This was the greatest killing ground, as the motorized rifle regiment would expose itself here on all sides. Speed, however, would probably get the enemy through here with sufficient remaining strength to be a serious threat to a successful defense. Accordingly, Evans would have to pull back from Battle Position 1 (BP 1) to BP 7. Baker would have to pull his three platoons from BP 2 to BP 6. While this shift was going on, the second belt of defense would contribute its arms to killing the enemy—Team Alpha fighting from BP 5 in front of Hill 910 and Team C fighting from BP 4 (with two tank platoons) and BP 3 (with one Bradley platoon).

Mines, wire, and tank ditches would thicken the defense in front of the engagement areas. The infantrymen from A, B, and C would have to move close enough to the obstacles to cover them, thereby detaching themselves from the subsequent movement of their parent companies to the alternate battle positions. This worried Always, yet he could see no alternative other than to cancel the movement, a sacrifice he could not afford.

Team C under Captain Carter would remain in its central position at BP 4, shifting the orientation of its fires to the rear as the majority of the enemy came by him. During this period of time, hopefully, the repositioned Echo and Bravo elements would add their fires to the battle. The aviation's attack helicopters would move up along Axis NORTH, thereby forming a gigantic kill sack in the middle of the sector. The final fight would occur in front of Team Coving in BP 11, Team Delta in BP 12, and a relocated Team Alpha in BP 8. The terrain at this point would restrict the enemy's movement, and if his attrition had been great enough in the earlier stages of the fighting, sufficient defensive firepower and the final belt of obstacles should stop him cold. At least, that was the hope of the defending task force.

Always was optimistic. The artillery officer had presented an excellent plan that covered the map with scores of targets. The air force officer promised no less than eight sorties to add to the close-in fires of the ground defenses. Technological advances in the ordnance guaranteed the air force would destroy whatever it hit, and probabilities were high that the artillery would stop a significant percentage of the vehicles and virtually all infantrymen caught in the open. Accordingly, the mortars initially would be in direct support of Team Bravo, then chopped to Team Charlie as the fight broke at PL MIDFIELD. The enemy had air to help him, but this time Always could preposition his air defense assets, avoiding the confusion that had come with movement in the previous two battles. The air defense platoon broke into two sections, covering the airspace between FORWARD and MIDFIELD, and between MIDFIELD and STOP. The Stinger gunners were assigned to A, B, C, D, and Team Coving, dug in and connected to the company commanders they were supporting with field phones. They would be ready and waiting when the enemy aircraft appeared, even if they failed to get any advanced warning. To negate that possibility, Always arranged to have their platoon sergeant stationed at the TOC, where he could monitor the air defense early warning net.

The scout platoon was broken up into squads, and squads were further broken down into mounted and dismounted teams. The dismounts would take to the high ground, monitoring any movements and giving advance notice of major developments (for example, "the main attack is coming in the north," "the lead battalion is caught in the mine field in front of BP 8"). The mounted teams would cover all routes into the sector, looking in particular for the enemy mounted reconnaissance effort that would likely occur that night, and giving early warning of the enemy main efforts during the day. Their mission was to avoid contact until the enemy had committed himself, then to continue to report the foe's activities throughout the depth of the battle.

The engineer and artillery staff had coordinated the placement of artillery-delivered mine fields. These would be on call to plug any holes in the defense and stop the enemy dead in his tracks. They would complement the obstacles already in place at the start of the battle and, as in every other case, would be covered by direct fire.

To the task force commander, the plan looked complete. Moreover, it appeared to be understood by his key subordinates. He had required each company commander to back-brief him on his understanding of the concept; the mistakes had been few and quickly corrected. As insurance Always scheduled a visit to each of his company command posts later in the day for a final review of company plans. Always would leave nothing to chance this time. The orders group broke up in an optimistic mood. A prayer had been given by the battalion chaplain asking for their success (interestingly enough, there are chaplains in Purgatory), punctuated by a fiery closing comment from Always to the effect of let the bastards come! Six and a half hours remained until darkness.

By midafternoon the enemy had a fairly clear picture of Always' intentions. He perceived that the defenders had spread out their forces over the breadth and depth of the battlefield but had failed to cover the extreme right flank of the sector. There was no room for an attack there, but just south of BP 3 was a cut in the ridge line that offered an opportunity for the scouts to infiltrate through that night.

Throughout the day the enemy was able to call in artillery fire on the defenders, seriously impeding their work efforts. The worst of these fires fell on Bravo Company in BP 2 and Echo in BP 1. Each company suffered several casualties during the day and had to take cover, losing precious time for digging in. At last light a spoiling attack momentarily disrupted Baker on

BP 2, costing him two vehicles and four infantrymen. The enemy suffered heavier losses, but he possessed the ability to replace them in time for the morning's attack.

Perhaps even more damaging was the enemy's ability to pick out the TOC location as well as the positioning of Always' obstacles. Throughout the day this information was passed back to the enemy commander, who in conjunction with his staff worked up a plan of attack that would avoid the strength of the defenses. In order to slow down the emplacement of the obstacles, he occasionally called for harassing artillery fires on the infantrymen emplacing them. He scored a major success with a direct hit on one of the bulldozers, killing the operator and puncturing the transmission. This exacerbated the time crunch that the defenders were already facing.

By dark the work effort was only about 40 percent completed. The obstacles forward had been emplaced, but the bulldozers had not been released to the waiting companies to the rear. Bravo and Echo were reluctant to let them go until they had dug in all of their vehicles, and Captain Baker insisted they also dig holes for the mortars attached to him. This caused great confusion when after dark the section sergeant was unable to locate Charlie and Alpha companies. After four hours of drifting (and quick snatches of sleep by the exhausted operators), they were located by Major Rogers and put back to work. They had only begun to emplace the obstacles in front of BP 8 and BP 11 when dawn brought the enemy attack.

Nor did all the rehearsals get completed. Bravo and Echo had been distracted by the artillery and spoiling attacks during the day and had never made it back to BP 6 and BP 7. After dark Evans and Baker were able to rehearse their own movements, but since their platoons were already scrambling to make up for lost time, they left them to continue their work. The competing pulls on the engineer officer, who had to oversee the obstacle effort as well as the leadership of his own platoon, had kept him from meeting with the artillery officer on the ground to plot the exact grid coordinates for the indirect fires covering the obstacles. The time lost in this delay also kept the artillery officer from making his way to Delta and Bravo companies to coordinate the fire plan with their respective artillery lieutenants. As a result, the two young men were not sure at what point they would take over the requirement of calling for fires in the face of the attack and what targets were their specific responsibility. This lack of understanding was never fully expressed to their company commanders, who therefore failed to bring it up with Always when he visited them after dark.

The battalion leaders were active throughout the long day and night, and wherever they discovered a snafu they rapidly put things back together. Every soldier, from highest to lowest in rank, was committed to a vigorous defense. There was no lack of desire on anyone's part, and it would seem that with a full understanding of the commander's intentions the mission could not fail. But the execution of a plan is difficult. The forces of nature, human and environmental, often combine to make even the most simple things difficult. Jeeps get flat tires, messages are misunderstood, units are diverted by uncoordinated leaders, soldiers get lost, winds blow down an antenna, map markings get smeared and are redrawn a few hundred meters off the intended location, weapons malfunction, transmissions overheat, and vehicle loads fail to get to their intended recipients. The list of calamities, great and small, is endless. Yet with strong leadership and aggressive initiative things are set right for a little while at least, only to be caught up again in another wave of confusion.

Always was gaining a keen appreciation of the complexity of leading a heavy task force. Just getting to visit all of his units, spread out over so great an expanse, was a chore. He reflected on the cavalier manner in which he had belittled such units in his lifetime, and how he and his buddies had laughed derisively when they recounted a tale of some botched-up battalion tripping over itself trying to complete a mission. How easy it had been to sit comfortably removed from the fray and critique the failings of others. How difficult to have the mission yourself. How pompous to spout doctrine knowingly with an attitude that success lies in the knowing, not in the doing. How humbling to discover the excruciating difficulty that lies in the doing!

It was shortly after 2200 that he returned to the TOC, sore and tired, to get an update from his staff on the readiness of the battalion for the morning's effort. Stoically he listened to the reports of enemy vehicular movement working its way in from the south. He was glad to hear that defenders at BP 3 had caught and destroyed two BRDMs (Soviet reconnaissance vehicles) slipping in behind Hill 899 at 2100, but he was concerned that they might have missed a couple that tried to come through earlier. He glowered at his S-2, probably unfairly, for overlooking the avenue of approach to the south.

Replacements were still coming in from the field trains, and he hoped they would be able to link up with their squads and vehicles in the dark. He was disconcerted to learn that one of the fuel tankers had gotten lost in the dark (there were not enough night vision goggles to go around) and blundered into a mine field. He considered restricting further movement at night, but dismissed the thought when he remembered how much work remained to be done. The scouts were reporting all types of movement but had been unable to put together a coherent picture. So much was going on that it was difficult to differentiate friendly movements from enemy activity, particularly when the distances were so great. Always could see that this would be a trying night.

He called his forward companies and warned them to be on the lookout for enemy probes into their sectors, particularly in and around the mine fields. It was imperative that the obstacles be in place by morning. When Bravo Company got into a major firefight at 0100, Always was glad he had urged vigilance. Mortar and artillery fire with variable time (VT) fuses had chopped up the enemy infantry, leaving them with two platoons of dead and wounded. Captain Baker immediately set to work to repair any breaches in the existing mine fields.

Always himself had slept between 2300 and 0200, awakened for thirty minutes during the fight at 0100. Despite his exhaustion, his sleep was not deep, and when he moved back into the TOC for a final update before leaving for Charlie Company he was having trouble focusing on the reports from the assistant S-3 and the S-2. He left instructions for a net call at 0300 and departed for his forward command post, located in the center of the defensive sector and north of BP 4. Major Rogers, realizing he could not observe adequately from the TOC, had already moved up in an armored personnel carrier to take his post on the right with Bravo Company; Major Walters was making final arrangements with the combat trains in regard to medical evacuation and was due at the TOC at 0300.

The big Bradley vehicle moved with surprising quiet across the desert floor. The sky was clear, full of stars that lent a serenity to the night air. Lieutenant Colonel Always relaxed, allowing his crew to do the lion's share of maneuvering the fighting vehicle into position. The driver picked the way in the dark, helped by his night vision goggles, the gunner keeping a vigilant lookout for encroaching enemy. A few minutes before 0300 they crept into place beside Charlie Company, the artillery officer and the air force liaison officer close behind in their M113. For a moment Always allowed himself to enjoy the pleasant coolness of the early morning, then set to work memorizing the call signs that he would be using at a rapid-fire pace within the next hour.

The net call went well, all parties coming up and sharing information on work completed, unit strengths, and enemy sighted. The intelligence officer had the hardest task, having to make a best guess on which avenue of approach the enemy would use for his main attack. In fairness, it was virtually impossible to predict with any degree of certainty. The desert floor, being wide open at the forward edge of the battle area and extending so deep into the rear, gave the enemy ample opportunity to shift his direction of movement. Having to make a call, the S-2 chose the northern avenue as the most likely, hitting Echo Company in BP 1, then swinging north of Alpha at BP 5 before picking up the wadi that hugged the south face of the mountain range to the north. Always agreed that this seemed most likely but understood that anything could happen; he tried to keep an open mind about where the action was likely to break. He sought to instill that sense of flexibility in the minds of his commanders.

An enemy reconnaissance outpost element of two scouts had been picked up by Alpha Company atop Hill 910, but although questioned they either had no knowledge of or refused to reveal the enemy plan. Always' own scouts had penetrated to the high ground forward of the battle area and so should be able to give him early warning of the direction of movement of the enemy. The radios checked, the intelligence shared, and the battalion brought to full alert, it was time to stand by for the attack.

Save the slight lightening of the sky in the east, 0400 came and passed without incident. Minutes ticked by slowly as the battalion strained to stay alert in the peaceful half-light. A morning haze was forming on the desert floor, partially obscuring the prominent terrain features so they could not be easily identified in the poor light.

"Bravo 36, this is Sierra 18." It was the forwardmost scout calling to Always.

"This is 36." The colonel looked at his watch. It was 0419.

"Bravo 36, this is Sierra 18. I can see smoke rising up over the haze beneath my position on Checkpoint 44."

Always looked at his map with a pen flashlight. Smoke was going in beneath Hill 876, center of sector 4,000 meters west of PL FORWARD. A few minutes later other scout reports revealed that the smoke was settling in all across the front, with the wind taking it directly into the face of the task force. The battalion commander marveled at the enemy's capacity for smoke production. Where did it all come from?

By 0435 the haze and smoke had combined to obliterate the view of the battle area by Bravo and Echo companies and all the ground in between. The thermal sights were of no use; although they were able to see through smoke, the attenuation of the haze and smoke made recognition of hot spots—the basis of the image—extremely difficult.

"Bravo 36, this is Sierra 18, over."

"Three six. Send."

"This is Sierra 18. I can't see a thing, but I can hear a whole bunch of tracked vehicles moving to my north, over."

"Roger, keep trying to see something. I need to know how much and what direction. Out." Always noted that the first indications were that the S-2 had been right. He had the TOC check to ensure that everyone on the net had heard the scout report.

"Bravo 36, this is Echo 36. My forward elements have spotted a line of BMPs and T-72s coming at them. We're taking them under fire. Over." Captain Evans's report was short and to the point.

"This is Bravo. Roger. Break. All stations this net, looks like enemy effort is coming in the north. Out."

In the moment during which Always transmitted his message, three events combined to hazard any further update on this assessment. The TOC, long since located by enemy reconnaissance, came under extremely accurate artillery fire mixing high explosive and persistent chemical munition; Charlie Company came under chemical and smoke attack (Always was into his gas mask in an instant and was immediately unable to see a thing); and the battalion radio net was jammed. While this was happening, the lead company of the attacking motorized rifle regiment broke to the south and headed directly toward Bravo Company at BP 2. The battalion of which it was a part immediately followed to the rear of BP 2.

Always cursed to himself. By this time it was apparent that at the critical moment he was likely to have minimum command and control. He did not yet know that his TOC had gone out, but he did know that he could not speak to anyone, could hardly see beyond twenty meters in front of his vehicle, and that the fight was about to break on his forward elements. He had known this was likely to happen, but that knowledge did not make him like it any better. He resolved to keep his own sense of balance while his TOC worked everybody up on the alternate frequency and he could reestablish communications.

The smoke in his immediate vicinity was incredibly heavy, as if the enemy had singled him out as a target. He dispatched Private First Class Davis, a soldier he had brought with him from the TOC for just such an eventuality, to signal the artillery officer's track that he was moving out. He then recovered the runner and shifted to the south in hopes of breaking free of the smoke. It was to no avail; more smoke descended to his front even as he moved.

In the meantime Captain Baker was making a valiant fight in and around Battle Position 2. The lead enemy company had run into the tank obstacle to Baker's immediate north and was virtually annihilated there, but not before it began to open a breach. Bravo Company was handicapped by the chemicals falling on its positions and by its jammed radio net. Nonetheless, Baker had organized a good fight and was making the enemy pay a price. Major Rogers was in the vicinity and recognized that, contrary to the last report from Always, a lot of enemy were piling into the south, attempting to bypass Bravo Company. As he tried to move to a better vantage point to assist in the fight and at the same time reach his battalion commander by radio, he was hit by an RPG round that glanced off his turret. The concussion and the subsequent swerve by his driver propelled his helmeted head into the side of the interior of the Bradley, knocking him out cold.

Baker was able to bring the second company under fire and hold them in the tank ditch, his dismounted infantrymen putting up a vicious fight in and around the obstacle. The third enemy company of the lead battalion, however, found a way around the obstacle, signaled the following battalion, and started a move to the east. The fight was going to come to a head at Battle Position 3, held by 1st Platoon of Charlie Company.

Echo and Alpha companies were frozen in place, not so much by enemy force (although they were taking a pounding from enemy artillery) as by the last report from their battalion commander that the main attack was coming at them. They strained to look through the lenses in the gas masks at the haze and smoke to their front. Echo Company was fixated further by the knowledge that it had killed at least two of the enemy vehicles originally sighted in the opening stages of the fight. Surely there would be more to come.

Major Walters had been wounded in the neck by the artillery crashing in and around the tactical operations center. He bled heavily, since he could not effectively treat the wound because of the gas mask he was forced to wear. He recognized the dire straits that communications were in, but was momentarily powerless to get a radio back up on the net. An enemy commando team had penetrated down the ridge line from the north and was putting the TOC area under direct small arms fire. With the casualties, smoke, chemicals, and artillery fire, the TOC could not get itself back into operation.

By this time Always recognized that he was getting no assistance in bringing commanders up on the net, and he set about establishing communications himself. Unfortunately this distracted his attention from directing his driver through the smoke.

Before they knew it they crashed into an unseen wadi that dropped five feet from the desert floor, leaving the rear of the Bradley hanging up a sheer wall, devoid of any traction and unable to extricate itself. Always jumped from the vehicle and began to move to his artillery officer's personnel carrier, only to see it get hit by an incoming artillery round, blowing its track off the rollers. At this point the battalion commander and his runner, the fire support officer, and the air force liaison officer began a cross-country dash to Charlie Company, hoping to find a vehicle from which they could command the battalion and coordinate their fires.

Captain Carter, by now up on the alternate net, had noted the lack of direction from above, and after attempting to raise the commander, executive officer, and S-3, concluded that command had devolved to him. At the same moment his 1st Platoon leader gave a call that an enemy battalion was crashing in on his position.

"All stations this net. All stations this net. This is Mike 36. I have assumed Bravo 36's duties. The enemy is attacking in the south. I say again, the enemy is attacking in the south. Acknowledge." Carter gave the call, the right one to make at this time, and received acknowledgments from Coving and Evans. Baker was too busy with his own fight, and Dilger had not yet reestablished communications. Archer was off the radio, supervising the action on the ground.

While Always and his small party made their way south, packing a PRC 77 radio that operated only in the clear, the defensive artillery battle was in the hands of the forward artillery officers. They were handicapped by the smoke and haze, but were able to call in the fire by occasional sightings and the roar of the enemy engines. Unfortunately, their timing was off. By the time they called for the preregistered fires, the enemy columns had bypassed the target areas being called in. The fire support was just missing, sparing the enemy a dozen or so vehicles he would have otherwise lost. As a result, the second enemy battalion crashed into Carter's 1st Platoon with full force. The fight was intense, and for a moment Lieutenant White, who had joined the battalion on the eve of its deployment, was able to stop the enemy with a wall of direct fire augmented by the mortar platoon directly supporting his company. But massed enemy forces eventually took out White's four vehicles, wounded him, and drove on to the west.

Now the third battalion of the enemy regiment was following close on the heels of the leading battalion in the south. It was into this battalion that Carter, leading his remaining two platoons, smashed in front of BP 3. For a second it seemed that the enemy would be stopped, but just at that moment air forces came in out of the sun and drove Carter to cover. Charlie Company's Stingers went into action and put down one of the jets, but more came on to keep Carter pinned down while the enemy drove on.

Captain Coving in the north was torn between Carter's call that the attack was coming in the south and Always' last report that the main attack was coming in the north. If he moved out of position, the enemy would have a free run across PL STOP. If he stayed put, Captain Dilger would have to make a stand by himself. He weighed the consequences, decided it was best not to leave his position uncovered, and prepared to defend where he was.

The battle had become sheer confusion for attacker and defender alike. Always had made it to Charlie Company's position by foot, only to discover that it had moved on. He pressed on to the sound of the fight. The enemy regimental commander had himself been put out of action, taking a direct hit from Lieutenant Wise just before Wise went down. The enemy's first battalion was still jammed up around Bravo Company, which by this time was seriously attrited. Smoke and gas hung over the battlefield. Soldiers fought by sheer instinct, feeling their way in the confusion, trying to kill the enemy while keeping from being killed. Time ticked by amazingly fast.

At 0620 Always made it to a disabled tank belonging to Charlie Company. After a quick exchange with the tank commander, the battalion commander took over the vehicle, which although affording him no mobility allowed him to talk on a secure radio. Just as he was getting updated by Captain Carter, elements of two battalions crashed into Captain Dilger at Battle Position 12.

Dilger had received last priority on the preparation of his defenses. The bet was that the main attack would go north against Coving, and the emphasis had been put on stopping the enemy forward, or at least depleting his forces in the early fight. As a result, Delta's obstacles were not as formidable as they might have been. Dilger had a three-platoon force (one Bradley and two tank platoons—given his maintenance status, a total of ten vehicles). Forty-three BMPs and T-72s threw themselves at the right-handmost platoon, consisting of three Bradley vehicles and fourteen dismounted infantrymen.

Always had a rough idea of what was happening from his tank up by Battle Position 3. It was do or die now, and he scrambled to call his helicopter support, which had been grounded until now by high winds, and to bring some air force into play. The latter had been hesitant to commit without any contact with the ground liaison officer. He also ordered Captain Coving to move his two-platoon force down to help Dilger. Coving tried to pick his way through the mine fields and obstacles to his south, but was handicapped by his lack of a rehearsal. Again the enemy jammed the battalion net off the air, and again the task force reestablished communications on an alternate net. By this time the battalion had been in gas masks for almost two hours.

For fifteen minutes the battle raged to the south of BP 12. Friendly air force arrived but was unable to pick up the enemy

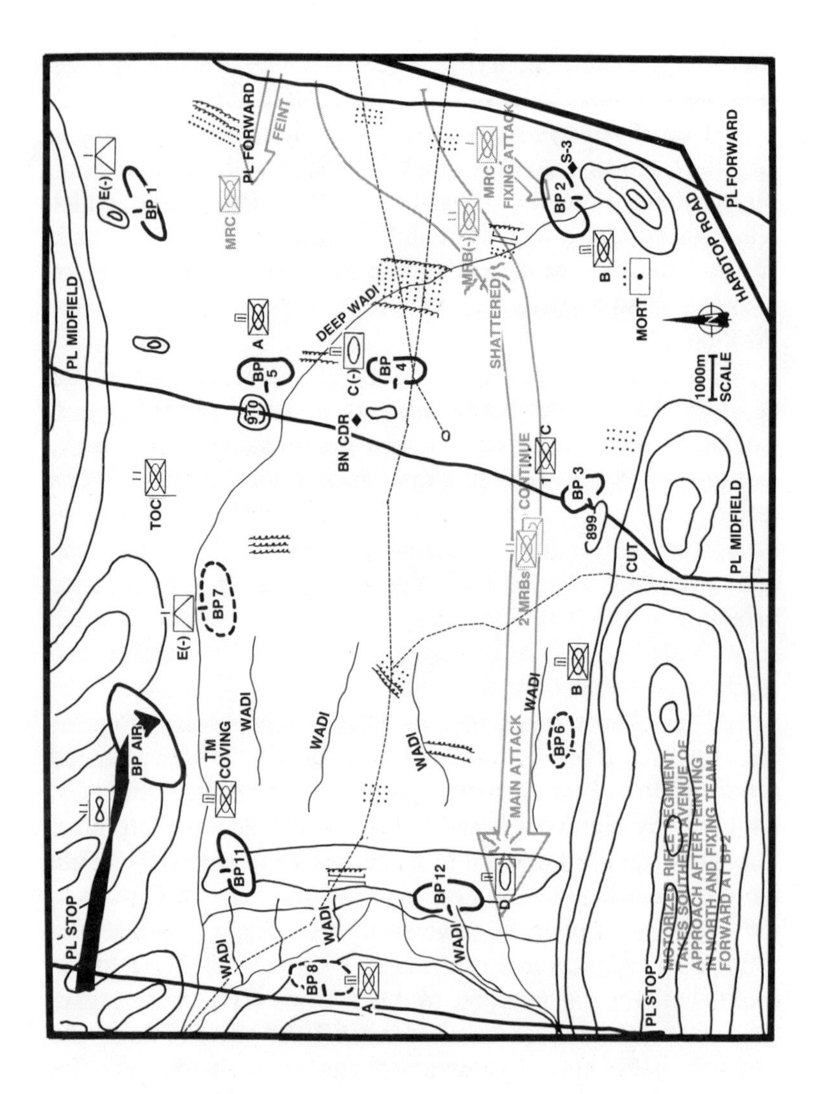
PL FORWARD
FEINT
MRC
E(-)
BP 1
PL MIDFIELD
DEEP WADI
A
BP 5
910
C(-)
BP 4
BN CDR
MRB(-)
SHATTERED
MRC
FIXING ATTACK
S-3
BP 2
B
MORT
HARDTOP ROAD
PL FORWARD
1000m
SCALE
TOC
C
CONTINUE
BP 3
899
CUT
PL MIDFIELD
E(-)
BP7
2 MRBs
B
WADI
MAIN ATTACK
BP6
TM COVING
WADI
WADI
WADI
BP AIR
BP 11
BP 12
D
WADI
WADI
WADI
BP 8
A
PL STOP
PL STOP
MOTORIZED RIFLE REGIMENT TAKES SOUTHERN AVENUE OF APPROACH AFTER FEINTING IN NORTH AND FIXING TEAM B FORWARD AT BP2

in the smoke, dust, and haze below, although it did manage to momentarily deter the enemy assault. Winds forced the helicopters back, but not before they picked off two BMPs, losing one chopper in the exchange. Finally, at 0703, elements of two enemy battalions, approximately thirty armored vehicles, crossed STOP. Always had once again failed in his mission.

Across the battlefield a bitter taste entered the mouths of everyone in the defeated battalion. Captain Baker felt no satisfaction with the two dozen enemy vehicles destroyed to his front. Lieutenant White suffered more from the knowledge that two battalions had run over him than from his wounds. Major Rogers' head was splitting from the pain of a concussion, but the ache in his heart was greater. Private First Class Davis, who had run three miles that morning with the command party, was close to tears at the news of PL STOP being crossed. From top to bottom, the ranks of the task force did not like, could not stand, defeat.

But in the seeds of the disappointment was the recognition that it had been a close run, that with just a little more luck, with a little more coordination, perhaps with a little more leadership, the enemy might have been stopped. The survivors sensed how close it had been, and being good soldiers in a good unit, resolved to rededicate their energies to the "next time," to learn from their mistakes and observations, and to wrench from defeat the fruits of victory.

Lieutenant Colonel Always recognized all of this and sensed that herein lay the key to solidifying his unit. A fever began to burn in him, a fever born in part of fatigue and frustration but born more of a zeal to win. The recriminations were over for Always. He had reached that stage where he was no longer the detached and aloof commander developing his command. The command and he had blended into one. The battalion was an extension of his will. As he willed himself to redouble his efforts, so too he willed the battalion. It would become an unstoppable force; as long as the fire burned in Always, it would burn in every one of his soldiers, a single force driving to the relentless objective of victory.

Lieutenant Colonel Drivon was merciless in the after-action review that took place at 1100. Always did not care. He was indifferent to the jibes, unbothered by the cataloging of his failures. His ego had been bruised beyond the point of feeling. But he did seize upon the kernels of truth, recognized what could have been done better, and branded the lessons learned into his own brain:

Defense is tough; not a minute can be wasted in the preparation. Make the plan early and get forces in motion. Revise the plan as a greater appreciation of the terrain and enemy intentions is achieved. Coordination is critical—engineers, artillery, air force, air defense, and maneuver forces must have the exact same plan in mind, on the exact same piece of ground.

Rehearse all movements, in gas masks, with no communications, and at night. Assume that confusion will reign at the moment of execution, and rehearse all procedures accordingly. Share the plan so that every last leader can assume responsibility for its implementation.

Time the movement of the enemy and adjust indirect fire accordingly. Anticipate the arrival at target grids and call for the fire on the spot to which they are heading, not where they are.

Use deceptive techniques in defense. Use smoke to cover obstacle preparations. Shift companies after dark; assume they have been sighted and reported by enemy reconnaissance. Commit heavily to finding, killing, and blinding enemy reconnaissance. Move the TOC before the fight and provide for an alternate TOC during the fight. Dig in all command and control vehicles that join the maneuver companies for the

fight. Put a ranking leader in charge of the bulldozers. Don't allow any deviations from their work schedule.

Preposition ammunition and protect it from indirect fire. A single platoon may make the critical fight against the bulk of the enemy forces. With well-prepared positions and ample ammunition that platoon can turn the battle.

Be bold and aggressive. An enemy seeking only to get past you can avoid a fight. Don't let him. Go to him and make him pay the price. Movement is important to the defense. Any uncommitted element must be prepared to fight as the reserve. In the absence of orders it must determine where the enemy is and go after him. Every resource must be brought to bear on the enemy; spare him nothing. Then and only then will you stop him, and that is what you must do.

With these lessons seared into each of their minds, the battalion operations group left the after-action review. Already an order was waiting for them. Always was eager to get it. He wanted another crack at the enemy.

CHAPTER 5

Deliberate Attack

The battalion would be attacking at dawn across the forward edge of the battle area, past Hill 876 and onto Hill 780. As usual, intelligence was sketchy, but Always had an advantage this time in that Lieutenant Wise and some of his scouts were deep into the enemy area. In fact, the outpost atop Hill 876 was ideally located to see the enemy dispositions. The commander had the S-2 signal Lieutenant Wise to get himself in position with all caution taken to prevent his being discovered by the enemy. Surely he would immediately expect a reconnaissance effort and commit forces to pick it up.

It was already past 1300 by the time the battalion leaders were free of the after-action review. Not much time was left to develop a plan, reconstitute the task force after the morning's fight, get elements in position, and mount the reconnaissance effort it would take to probe the soft spots of the enemy.

Major Rogers had passed a warning order immediately to all companies, and even before the commanders made it back to their respective units the company executive officers were cross loading the vehicles and platoons so that they would be ready to fight on short notice. The commanders themselves were dog tired, but they would have to put that aside in order to get their people in motion. For the moment they could concentrate on reconstitution, as it would be a while before Always and his staff would decide what the plan of action would be. They would need every minute they had to pull off an attack at dawn.

The colonel went forward to Bravo Company's position in Battle Position 2 hoping to gain a vantage point from which to look into the enemy's sector. With him was his S-3, the artillery officer, and the S-2. Captain Baker joined them for a few minutes, offering his views on the problem before them.

"The ground is wide open, sir. Hill 876 and the ridge line marking the northern ring on the hidden valley across from my position are the dominant pieces of terrain forward in their sector. It seems like any approach along the south is sure to be caught in a murderous cross fire."

"You're probably right, Captain Baker," Always answered him, "especially considering that peanut-shaped hill to the south of 876. Yet it is the shortest approach, particularly if we use your position here as our last cover before we launch."

For a few minutes the party shared ideas. Lieutenant Colonel Always asked to see the operations overlay passed to him from Brigade. The northern boundary of his zone of attack seemed to close out any opportunity of using the ground up there to avoid 876. The zone was too narrow and pinched him in against the dangerous ground where he was sure he would find the enemy. Yet in its very unattractiveness he saw an opportunity. "Let's work our way north along the FEBA. I want to take a closer look at those two little knobs that jut out of the desert floor just north of Hill 876. Major Rogers, call Brigade and see what kind of latitude we have on moving our left boundary."

The small party worked their way cautiously along the FEBA (forward edge of battle area), having to cut back toward Hill 910 before they could move in on Echo Company's position. The enemy was able to observe deep into Always' sector, and the colonel saw no sense in drawing artillery fire. The last thing he needed to have happen now was to lose one of his key men or to strap the command group with a casualty or two.

Brigade was firm. There would be no alteration of the boundary. Using his binoculars and map, however, Always was able to see what he was looking for. He reflected for a moment or two, then bounced his ideas off the command party.

"It seems to me that the enemy has got to expect us to attack in the south. While he doesn't know exactly where our boundaries are, he can probably figure them out with good approximation. If he does that, then he will deduce that without cover in the north I can't mount an attack there."

"Well, he probably has at least one company in defense, and probably two or more." It was the S-2 speaking. "That gives him enough forces to hedge both bets."

"You're right on that, but it's not his first line of defense that worries me as much as his reserves. If I can delude him long enough to bust through his line, then maybe I can get deep before he can react."

Always had looked hard at the map and determined that getting to Hill 780 would unhinge any defense by the enemy, provided he could get there with adequate forces. "Where do you think he'll put his reserve?"

Major Rogers fielded the question. "Can't tell for sure, sir. Maybe our scouts can give us a better picture over time. It's most likely, though, that he'll put them in that valley in the south of our zone."

"I think so too," the S-2 chimed in excitedly. "Although it's not a great place to put them, he does have to worry about us coming into that valley, and it offers him some cover from artillery fire and observation as well. So, like as not, that's where he'll put them. If he does that, then he's got a problem coming out of it, since the only exit is looking into the face of Hill 780."

"What can we do about that with artillery?" Always was looking at his fire support officer.

"Sir, if there ever was a place to put in an artillery-delivered mine field, that's the place for it."

"And how long would it take you to get it in?"

"If the guns were preset and laid, slightly more than five minutes, running up to ten."

"And if they were in action, firing other fire missions?"

"Then a few minutes longer, sir."

The conversation continued in this vein for twenty minutes, during which the essence of the plan was sketched out. That done, the party departed for the TOC, now reestablished in the vicinity of Charlie Company's position of the morning. Already it was past 1530 and much coordination had yet to be done. The order would be given after dark.

Always had been learning, however, and he remained in place while he called his commanders up to him. They would get the order later at the TOC, but now while it was still light he wanted them to see the ground over which they would attack in the morning. As precious as their time was, this was as wise an investment of it as any. Shortly after 1700 the assembled commanders returned to their companies. The orders group would meet at 1930.

On the way back to the TOC Always ran into Command Sergeant Major Hope, who was talking to two soldiers recovering mines from one of the obstacles emplaced the previous day.

"Good evening, sir." Hope greeted the colonel for his small group.

"Evening Sergeant Major, evening men. How are you all making out?"

"Just fine, sir," one of the men responded as he gingerly put aside an antitank mine he had just defused.

The sun had baked the men to a dark hue. Chapped skin hung from their lips and noses. Eyes looked as if road maps had been drawn on them, the red lines careening off in every direction. Always noted that their hands were raw and blackened by heavy labor in the arduous climate. Yet they looked cheerful enough, as if they were making picnic lunches instead of picking mines out of the ground. The sergeant major had that effect throughout the battalion, raising spirits even as the soldiers redoubled their efforts to get the job done. He came over to talk with his commander for a few minutes.

"How are you doing, sir? You look awful."

"Thank you, Command Sergeant Major, you're looking great yourself." Always smiled.

"We almost had them this morning, you know, Colonel."

"Yeah, I know. But that's not good enough. Tomorrow is another day. I think we're ready to really do it now." The smile on Always' face had turned to a set grimace. "How are the men looking to you?"

"They're ready, sir. They want to kick butt, and they feel that you can make it happen. Show them the way, and they'll knock walls down getting there."

Again the sergeant major had given his commander the lift he needed in precisely the right way. He was gladdened by the expression of solidarity and support, challenged by the expectations of victory, and resolved to deliver to his command the success they deserved. All this in a few words, a few casual gestures. No wonder the men of the battalion responded to Hope the way they did. He could move mountains.

For an hour Always brainstormed the plan with his staff. Ideas were exchanged, critiqued, restated, critiqued again. All the while reports were coming in from the scouts, filling in the details on terrain, and just as importantly, on the enemy.

Lieutenant Wise had gone to ground atop Hill 876. The enemy had committed the better part of a dismounted company to look for him and his men, and 876 had been an obvious spot. For two hours Wise and two of his men had buried themselves with dirt, sand, and rocks, allowing the patrols to pass within a few meters of them without their being detected. It had taken amazing stamina; the heat of the day and the stifling weight of the earth on their entombed bodies had severely drained them of essential fluids. But they had hung on and were now gaining a clear picture of the enemy sector.

A motorized rifle company, reinforced with tanks and a dismounted element of more than fifty men, was digging in around the base of 876, the two hillocks to the north, and the peanut-shaped ridge to the south. Along the north face of the ridge dominated by Hill 955 another motorized rifle company and two platoons of dismounts were preparing a gauntlet defense, ready to rip apart anything that tried to run it. There had been movement in Hidden Valley, but as yet no scout was in position to get a good count of how many enemy were there. Based on the noise and the dust raised, it seemed like tanks were in there, at least a platoon and maybe a company. Hill 780 was uncovered save for a few support elements coming and going. Obstacles were in everywhere; across the width of the sector from north to south stretched concertina wire, barbed wire, and an elaborate mine field. Another more intensive mine field stretched from the peanut-shaped hill to the ridge line in the south. The entrance to Hidden Valley was a maze of tank traps, mines, wire, and trenches. Each enemy vehicle was dug in to turret level, and the approaches to them were wired and mined. The road south of 876 was cut with wire and mines and covered by cross fires from both sides of the road. So far as the scouts could see, there was not an inch of ground uncovered by fire once an attacker came within range.

It was a bleak picture, but at least it was a picture. As bad as it was, for the first time the task force knew what it was up against. The planners solidified their views, Always approved the plan, and the TOC produced the overlays that would be discussed and passed out at the orders briefing.

The plan was elegant in its simplicity, electric in its violence. The overwhelming mass of the task force would strike in the northernmost edge of the zone of attack, set up by a small deception in the south, and assisted by a preparatory dismounted attack. The commanders' eyes brightened as they leaned forward in the eerie light of the TOC to be sure they understood what was expected of them.

Echo Company would move to Checkpoint 8 after midnight, run its vehicles' engines, keep some movement going throughout the early morning hours, and then make a false start forward at 0355, going no farther than the north-south road to its front. It would be feinting at Hidden Valley, hopefully fixing the enemy there while the extreme violence was done in the north.

Bravo Company's infantrymen would start out at 0130 to push at the northern slope of Hill 955, specifically aiming toward CP 4. It was to move by stealth, breaking into a direct attack at 0400. It would reinforce Echo's deception, further confusing and fixing the enemy.

Alpha Company's infantrymen would start out at 0100 from the vicinity of CP 5 and move by stealth just short of Objective OWL. Under cover of an artillery preparation that would start at 0330, they were to open a breach in the wire and mines in the extreme north of the zone of attack, just to the northeast of OWL.

Through this breach would come the rest of the task force. Alpha Company would drive in on OWL's north face, gathering in its infantrymen and reducing the defenses there, fighting position by fighting position. Seconds behind Alpha would come Bravo, minus its dismounted infantrymen, hooking behind OWL and cutting sharply south into Objective FALCON. Its mission was to engage directly the defenses of FALCON, thereby protecting Charlie's tanks as they followed right behind, hooking in from the north around OWL, past FALCON, and straight at

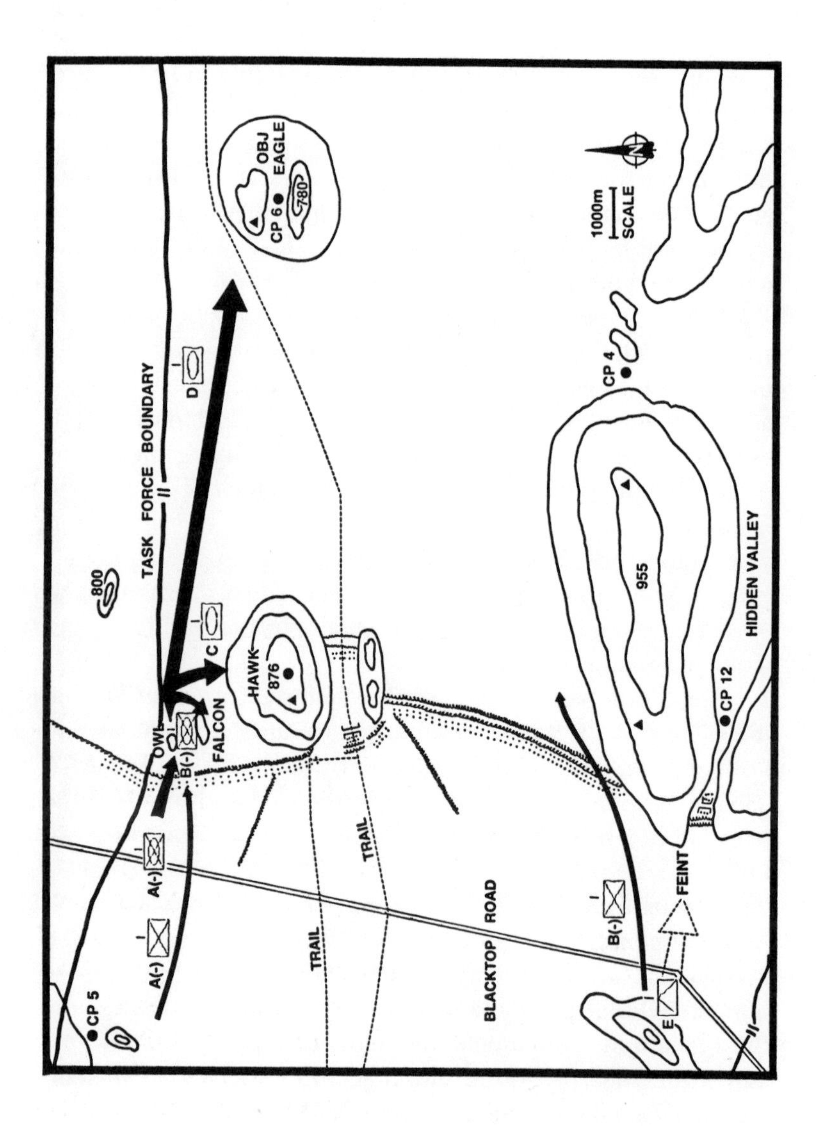
TASK FORCE BOUNDARY
OBJ EAGLE
CP 6
780
1000m
SCALE
800
D
C
HAWK
876
OWL
FALCON
B(-)
A(-)
A(-)
CP 5
TRAIL
TRAIL
BLACKTOP ROAD
B(-)
FEINT
E
CP 4
955
CP 12
HIDDEN VALLEY

Map 4. Attack on Objective EAGLE

Feint on Hidden Valley:
Company E: 3 ITV platoons (feint only)
Company B: 3 infantry platoons; no Bradleys (attack in south)
Attack on OWL:
Company A: 3 Bradley platoons; 3 infantry platoons
Attack on FALCON:
3 Bradley platoons; no infantry originally. Later Company A reinforces.
Attack on HAWK:
Company C: 3 tank platoons
Attack on EAGLE:
Company D: 3 tank platoons

Notes:
Chemical Platoon (smoke) attached to main attack.
Infantry to dislodge dug-in enemy armor once a toehold is reached on OWL, FALCON and HAWK.

HAWK. It was to grab a toehold on the north side of 876, covering any enemy movement that might interfere with Delta Company's thrust on Objective EAGLE. The idea was to get deep quickly with a tank company on EAGLE, blocking any counterattack attempts from Hidden Valley or from the east. The forces on OWL, FALCON, and HAWK would take their time to work the enemy out of their holes, the lion's share of this work to be done by the dismounts of Alpha coming down from OWL, and from Bravo, working up from the south.

The main attack would break at 0400, led by two armored personnel carriers with smoke generators on board. Visibility would be virtually nil for the attacking column, but with the task force covered all the way in by the poor light and spewing smoke, the few seconds needed to get by OWL might be bought by the confusion. Not that Always did not anticipate that his own people would be disoriented in the melee. That was to be expected. But with boldness and determination they could make it into position before the enemy could react, and then hunker down to fight a deliberate attack from positions less vulnerable than coming across the desert floor. Artillery-delivered mines would be dropped on CP 4 at the critical moment, bottling up whatever forces were in Hidden Valley, now estimated to be approximately ten tanks and a platoon of BMPs.

The bad news was that no air support would be available for the attack. It was needed elsewhere. Nor would there be any helicopters available. The rigors of the previous day's efforts in the high winds had caused several maintenance problems, and those aircraft that were available were committed further to the north. This would be strictly a ground attack, supported by artillery and mortars. The latter would go in mounted behind Alpha, supporting the infantry efforts in the north.

It had taken only a short while to describe the relatively simple plan. But its implementation involved a myriad of details, particularly for the night movements that would have to be made to get everyone in attack position. To the greatest extent possible, all movement would be done with radio silence. Deception was key. If the enemy realized the size of the punch coming in the north, he could reposition himself in time to meet it and throw it back. The scouts had been alerted to report any large night movements in and around OWL, FALCON, and HAWK. A scout team was pushed deeper onto EAGLE to make sure that no surprises awaited Delta when it got there.

The commanders were eager. This seemed like a workable plan, one based on good knowledge of the enemy positions. Come what may, it was certain to be exciting. The heart of the battle would be fought in the first ten minutes. After that it would take yeoman's work to uproot the enemy, but the clash of armor at 0400 would be a scene to remember.

At 2100 the meeting broke up. Always shared a few words

with Major Walters, whose neck wound was now bandaged, and headed for his Bradley. Shortly after midnight he would start his move toward CP 5, but for now he was going to get some rest.

As Always slept, the work of the battalion went on. Orders were developed and passed down the chain. Maintenance crews worked feverishly to complete repairs and get equipment back into the fight. Rations were distributed, water was brought up, ammunition was stored, fuel was replenished, mines and obstacles were recovered and stowed for future use. Here and there men snatched short naps. The leaders and the led had become one, caught up in the furor of continuous battle, straining to keep their gear in order, their bodies conditioned for the next fray regardless of the wear and tear they had already been through. One thousand wills bent to the tasks at hand, bone tiredness compensated for by the hard-won experience that comes with the tests of fire and steel that they had been through. Fear was subordinated to mission, exhaustion to commitment. The battalion had passed from being in the field to being part of the field. The comforts of civilization had passed from recent memory. Their bodies had hardened with their wills, and they could not be broken, come what may. They were now a machine that could fight indefinitely.

At thirty minutes after midnight Always had positioned himself astride the route into the attack position in the vicinity of CP 5. Straining his eyes to look into the darkness, he could pick out the units slowly making their way in. Alpha came first, silently disgorging its face-blackened infantrymen under the leadership of the three platoon leaders—two lieutenants and a sergeant first class. They quickly hoisted their packs and moved out, squad leaders counting their men by feel, checking for any equipment oversights as they did so.

Bravo followed, minus their infantrymen already in position down at Checkpoint 8. Lieutenant Franklin had brought up the

Bradleys, Captain Baker making the decision to go with the dismounts. Always caught himself second-guessing the call. Certainly Baker was weighing the difficulty of the two missions—dismounted and mounted. The infantrymen had the longer haul to make, the more enemy to work their way through. It was certain to be a difficult chore, requiring strong leadership. But the Bradleys of Bravo were key to the protection of the tanks. Moreover, Franklin had not been forward with the commanders when they initially looked at the zone into which they would be attacking. Nor had he been present in the TOC when the order had been briefed. Of the two missions, the one more important to the battalion was the attack on FALCON by the infantry vehicles. For a moment Always was torn between the urge to dictate that the commander must go with the main attack and the desire to let his subordinates make key decisions affecting their own commands. Then he realized the issue was moot. Franklin was here and Baker had already committed himself to the night movement. Franklin would have to take in the Bradleys.

Always wished he knew the lieutenant a little better. He looked tough enough; a short, muscular man, he had been an outstanding college wrestler. The colonel decided to put himself behind Bravo in the attack.

Over the next hour and a half the remainder of the battalion closed, vehicles moving slowly to muffle the noises of their huge engines. One by one Always saw them come in and disperse. Despite the losses from the defense in sector the previous day, he had enough to mount the attack. Twenty Bradleys and nineteen Abrams tanks would be going in at dawn. It was a minor miracle pulled off by his executive officer and the maintenance crews. More than ten percent of the task force was committed to that effort, more than a hundred men. No group worked harder.

The smoke platoon leader was excited. He had never taken his generators into an attack of this nature before. The prospect of leading a heavy task force into an objective both thrilled

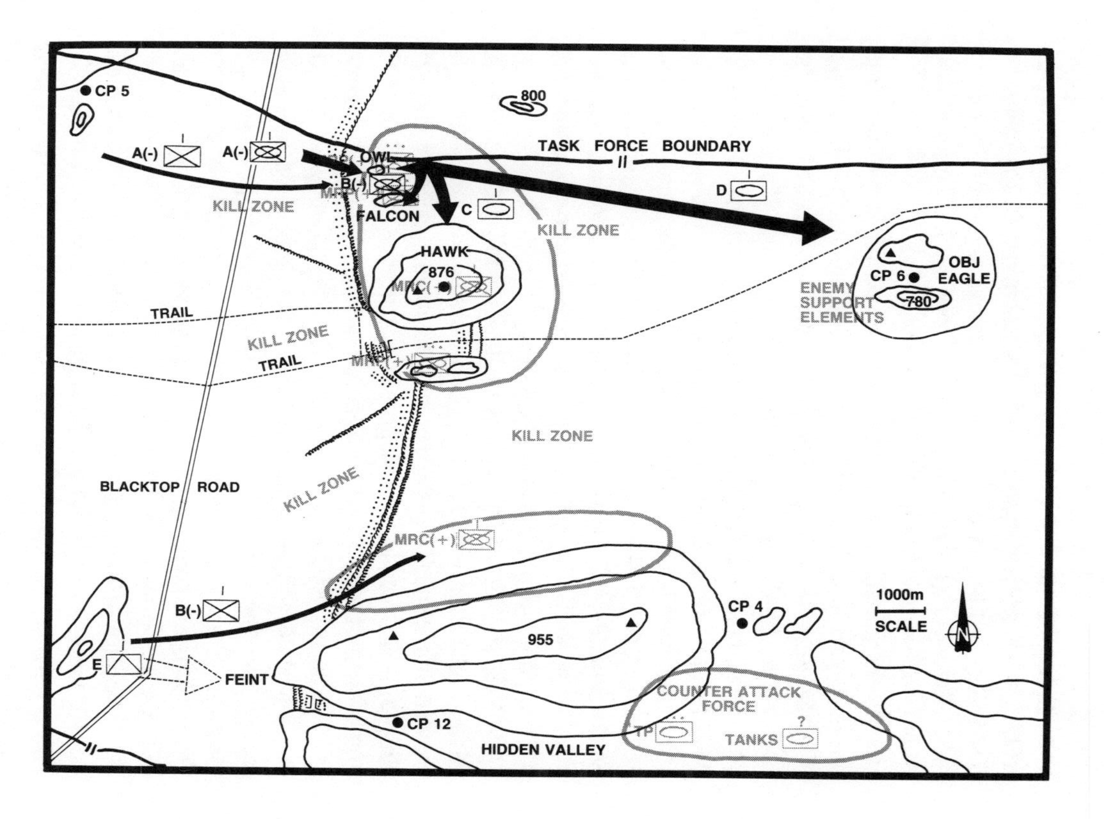
CP 5
800
A(-)
A(-)
OWL
B(-)
TASK FORCE BOUNDARY
D
KILL ZONE
FALCON
C
KILL ZONE
HAWK
876
OBJ
EAGLE
CP 6
780
ENEMY
SUPPORT
ELEMENTS
TRAIL
KILL ZONE
TRAIL
KILL ZONE
BLACKTOP ROAD
KILL ZONE
MRC(+)
B(-)
CP 4
1000m
SCALE
955
E
FEINT
COUNTER ATTACK
FORCE
CP 12
TP
?
TANKS
HIDDEN VALLEY

and frightened him. Lieutenant Rizzo was a good officer, an ROTC graduate who had chosen the Chemical Corps based on his college studies as a chemist. He always wondered if he had made the right choice, somewhat envious of his combat arms friends. This morning he was sure he had.

The lieutenant and the colonel were staring off to the east, trying to pick out the shadow of Hill 876 against the sky. "Set your azimuth relative to that peak when you start," Always was telling him, "and stick to it. Every time you can get a view of it, reset the azimuth. Your compass will be going crazy with the jolting and the pull of the metal on the vehicle, but it's better than nothing. Go as fast as you can, and once you get to the wire, let these guys come by you. I'll have infantrymen on the ground guiding them through with flares and green smoke. Keep that smoke coming—it's not only your best protection, it's the only cover I'll have as we go in."

"Yes, sir!" The butterflies were flapping in Rizzo's stomach. Always looked at him and knew he would do the job. A thought crossed his mind. We raise a boy to manhood, give him an education that leads to a degree in chemistry, and then send him into the face of hell as a smoke screen. What a business.

He patted Rizzo on the shoulder. "Keep up your speed. We'll be following the thickest part of the smoke, each vehicle racing after the one to its front, but we'll keep up. If you get disoriented, keep driving on until you can pick up your direction again. Whatever you do, don't back up or thirty-nine vehicles will run you over."

"Don't worry, sir. I won't let you down."

"I know. That's why I picked you. Good luck, Tony." Always made one of his rare uses of a first name. It seemed appropriate at the time.

The artillery belched, shattering the early morning tranquility. It was 0330. The TOC, having shifted position since midnight, ran through the radio net, quickly checking to see if all elements were in contact. That done, the radio frequency was shifted by prearranged signal, and a radio was left operating on the old net, chattering with a dummy station in the south.

Evans had been making noise over by CP 8 for more than two hours now. He drew some artillery for his trouble, but no casualties. He didn't like missing the fight, but knew that he had to play his role well. He was a professional. At the appointed time he started his platoons on their move to the east. The enemy saw him coming and passed the word over the radio net.

The infantrymen of Bravo Company had made their way quietly through the dark before going to ground about 800 meters short of the ridge line north of Hidden Valley. The point element had seen first one outpost and then another, and made its way back to Captain Baker, who wisely decided to hold up, not wishing to set off the attack prematurely. He positioned his men, and at 0355 moved in and quietly took out the two outposts. From here on in there would be no stealth. Baker's attack went in against the enemy infantry dug in along the north side of the ridge. By 0359 Baker had suffered his first casualties.

In the north, Alpha's infantrymen had found the wire. An enemy squad emplaced to overwatch it at the point of the breach had drifted off to sleep in the quiet of the night, blankets wrapped around their huddled forms. Quietly, Archer's men closed in on them. Under the cover of the noise of the first artillery shells, they shot them where they lay. Instantly they set to work to breach the mine field and cut through the wire.

It took no order to get Lieutenant Rizzo moving. As Always' wristwatch second hand swept over 0400, the smoke platoon pulled up out of its cover, smoke just beginning to rise from the back of the armored personnel carriers. In an instant the whole battalion was alive. An enormous din arose from Checkpoint 5. The earth shook from the almost forty steel monsters straining to accelerate to top speed from a standing start. Air defense, command and control, recovery, and maintenance vehicles mingled with the combatants, adding to the pandemonium. Somehow the units maintained their integrity, crowding together in the smoke, dust, and dark, racing the dawn's first light to Objective OWL. Vehicle commanders' faces were contorted with intensity, cool wind and searing sand biting into their skin, eyes shielded behind pitted goggles. Gunners forced their foreheads against their sights. Death was waiting. They had to see before they were seen. Life hung in the balance. Tank loaders braced themselves at their stations as best they could, whipping back and forth in the tumultuous ride. Any second they could swing into action, feeding the rounds into the smoke breeches. Drivers tried to keep sight of the vehicle to their front, appearing and disappearing from vision in the smoke, haze, and darkness.

Tony Rizzo was drawing fire. The smoke trailing behind him covered everyone else but left him naked to the enemy, shielded only by the weak light of predawn. He could see Hill 876 to his right front, checked his compass, and urged his driver to greater speeds. His young, strong eyes picked up the twin knobs of OWL and FALCON. Two thousand meters to go. Incoming rounds were kicking up the sand to his front. He swerved left, then right, then picked up his azimuth again, heading right at OWL.

It was working. It was working. Lieutenant Colonel Always felt the moment, saw the opportunity. Only a little luck. It would take only a little luck. He pressed up behind Bravo Company, switched its command net onto his second radio, and held on for dear life.

Rizzo saw the wire, turned slightly north, and picked out the flare. There was the breach. He had found the breach. The high explosive round tore through the side of his APC, sending spalding through his upper body. He slumped in his cupola, nothing left to hold his remains rigid. The second vehicle, commanded by the platoon sergeant, took up the lead and dashed for the breach. The battalion followed.

Always' artillery was pounding OWL and FALCON. It was enough to keep the defenders buttoned up, not enough to keep them from firing. Smoke rounds were mixed in with the high explosives. In a few minutes it would have to lift, but for a while longer it would help the attack.

The enemy was reporting the main attack to the south, causing the defenders there to brace themselves for the armor onslaught, and exaggerating their assessment of Bravo's infantry attack. The commander there was calling for commitment of the reserve as he fired at Echo Company in the distance, still out of direct fire range. The enemy forces on HAWK, splitting the distance between the northern and southern efforts, was not sure which way to orient their weapons. Coolly, they waited for developments.

Captain Archer rolled his Bradleys through the breach ana dashed for the north side of OWL. He was untouched as yet by the enemy fire. As he sped by he flashed a thumbs-up sign to his infantrymen. They had done a hell of a job, but their morning's work had just begun. The diesel fumes filled his nose and pumped a load of adrenaline into his veins. A BMP opened fire on him, and he triggered his smoke grenades and lunged his fully loaded Bradley into a depression 200 meters short of the objective. As the BMP inched up to get a shot, it was drilled by the platoon sergeant of 1st Platoon. Archer dashed the last 200 meters, splashing diesel across his engine, adding to the smoke screen under which his company advanced. At 0412 ten Bradleys of A Company were in on OWL. The infantrymen were sprinting in to link up.

Lieutenant Franklin sped past the second burning armored personnel carrier. A decapitated corpse stood in the commander's hatch as smoke continued to bellow from the generators in the rear. The sight shook him a little. Why was it standing? It had no head. It shouldn't be standing.

Always saw Franklin's track waver a bit, then bust through

the breach beside the destroyed platoon sergeant's smoke track. Eight Bradleys went through with him and Always followed. He thought he saw Archer's company hugging the north slope of OWL. That was a good sign.

Franklin continued to drive to the east.

"Turn! Turn! Turn right!" It was Always yelling to Franklin over B Company's command net. Damn, what was wrong with that lieutenant?

Franklin heard the call and wheeled his company to the right. He wasn't sure, but he thought he saw FALCON, his objective. "Let's go!" he yelled over the radio. He was heading straight at Hill 876.

"Oh, my God. No!" It was Always talking into the intercom. "He's missed FALCON."

"Franklin. To your right! To your right! FALCON'S to your right!" No call signs. It was Always yelling into Franklin's net.

"Right. I got it." Franklin didn't know who was talking to him. His eyes were fixed on HAWK, driving straight at it at forty miles per hour.

The enemy forces on the back side of FALCON had a perfect flanking position on Bravo. Under steady fire direction from their platoon leader they chewed up Franklin's column. Always turned his own vehicle into FALCON, spitting 25mm rounds from his main gun, trying to distract the murderous fire. "Franklin! Turn your company, turn your company, or you're done for." It was too late. Franklin was gone, and with him six of his vehicles. The seventh spotted Always and turned with him into FALCON.

The task force commander was now fighting for his life. He wanted to get a call to Charlie Company, to give a warning that FALCON was unsecure, but he would not live to complete the call if he did not get to cover. And the only cover available was straight into FALCON at the BMP trying to take him out.

"Spivey, bob and weave. Kick up some dust!" Always yelled into the intercom.

His driver responded instantly while Always yelled into Bravo's radio net to its sole survivor, "Come with me at the BMP. We've got to get his cover."

The buck sergeant commanding the vehicle understood the urgency of the command and opened up on the BMP as it was drawing a bead on Always. Like a well-drilled fire team the two Bradleys dashed into the small box canyon where the enemy lay, confusing him with their sporadic rushes, keeping him hull down. In a minute they closed the distance, and the Bravo vehicle put four rounds through the BMP. The two Bradleys went to the wall of the cliff at either end of the tiny canyon, huddled against the earth, seeking whatever protection it offered.

As he spoke into his mouthpiece to alert Captain Carter, Always saw Charlie Company turn the corner around OWL and head into HAWK. The attack could not be stopped. Direct fire opened up from FALCON and HAWK as Carter made a dash for cover. He was picked apart in the cross fire. Within a minute and a half his entire company of tanks was peppered. Two limped into the north face of Hill 876. The rest were burning on the desert floor, crews scrambling for cover from a few, the rest ominously silent.

The task force commander had watched two companies destroyed in the space of a few minutes. With FALCON unsecured, the movement onto HAWK had been suicidal. The breakthrough at OWL had gone smoothly. The plan had unraveled with Franklin's misorientation. Always now had less than a minute to decide whether to call the whole thing off or to press ahead with what he could salvage from the original plan. In less than two minutes Captain Dilger would be breaking into the open around OWL.

One minute to decide. One minute to decide. It all comes down to this. The success or failure of a mission—the lives of hundreds of men, the worth of the sacrifices of those already lost—is not debated with the best minds of the battalion present, with those whose lives are at stake, not considered with calm reflection and a coherent, logical analysis. Instead, it comes down to a moment's decision in the heat of battle, by a scarred and scared commander, personally fighting for his own life, carrying the burden of young lives already spent, trying to do his best by his men and his mission.

The thoughts crowded into Always' mind. There's no turning back now. Turning back is certain defeat. It's certain death. Delta is committed. They could stay at OWL with Alpha, but that sets everybody up for later defeat. The enemy will counterattack and throw us off our toehold. Rizzo will have died for nothing. Bravo will have been spent for naught. Charlie as well. Winning means taking EAGLE; beating back any counterattacks; giving Alpha time to work the enemy out of OWL, FALCON, and HAWK. The only chance lies in going for EAGLE.

"Dilger, watch your right flank. FALCON and HAWK are uncovered. Stay in my smoke. Take EAGLE!" Again the violation of radio procedure. But seconds counted, and Dilger got the message.

Always signaled to the Bravo Bradley with him to smoke and follow him. He then called the TOC to shift artillery into the north and west of 876, and launched his vehicle at EAGLE. Smoke trailed from the two Bradleys, leaving a screen behind which Dilger could hide the head of his column. Delta opened up with their smoke as well, elongating the screen. The wind whipped it to the north and east, uncovering the trailing platoon. At forty miles per hour the two Bradleys and ten Abrams careened across the desert.

The wild ride took four minutes. When it was over, seven tanks and one infantry vehicle held EAGLE. Three tanks and the Bradley from Bravo were burning along the route from OWL.

A few surviving crewmen were crawling for cover in whatever depressions they could find in the soft sand.

The crux of the fight was over. It was 0426. For four more hours the opponents would fight, Archer slowly and deliberately working his company around OWL, reducing enemy positions one by one, then moving on to FALCON, repeating the process before finally taking HAWK. The enemy deduced that the main effort was in the north and tried to move his forces from the south, only to have them stopped in Hidden Valley by artillery-delivered mines on Checkpoint 4 and withering fire from Dilger on EAGLE. A feeble counterattack from the east was stopped dead by the EAGLE force. Captain Baker kept the forces on the 955 ridge line pinned. Artillery sought, and found, infantrymen on either side. But the attrition was gradual. Eventually, Evans could enter the fight in the south, thickening Baker's fixing fire on the enemy. By 1030, the defenders pulled out what forces they could recover and withdrew south of Hill 955. The rest either surrendered or died in place.

Always had his victory, but it was an expensive one. The better part of two companies had been destroyed. The others had suffered casualties as well. Enemy air had entered the battle around 0900, and although they had been picked off by the air defenses that had closed in to OWL behind Archer, they had knocked out a couple of the remaining vehicles. Casualty treatment and evacuation had been extremely difficult, given the contesting of the terrain and the inability by either side to move freely around the battlefield. It was a victory all right, but a Pyrrhic one. Always was not content. It should have come cheaper.

The after-action review made the right points. The task force knew where the mistakes had been made. It had been a closely run thing, had turned on the missing of FALCON. The plan had been bold. It had almost paid off. Human error, always so common in warfare, had depreciated the success. This time Always found himself reflecting on those things that had gone right; he did not want to throw out the good with the bad:

Win the reconnaissance battle and you are well on the way to winning the fight. It doesn't take much, a few good men in the right place at the right time, with the presence of mind to know what to look for.

A deception, no matter how simple, pays off. Over great distances, in the confusion of battle, on the raw nerves of humans exposed to the terror of steel and fire, the obvious gets noticed and leads to hasty conclusions. Play the enemy against himself. Make him think what you want him to think.

Mass pays off. As expensive as the breakthrough had been, it would never have worked without the critical mass to continue to drive through. The combination that fixes parts of the enemy, screens out another, and pulverizes a third is a winner.

Coordination of a plan on the ground is achievable if the plan is simple enough, if elements are given enough time to get into position, and well-trained soldiers execute it.

Smoke cover is key. It complicates an attack, but confusion favors the bold if confusion is equally present on both sides.

The simpler the plan, the less radio chatter is necessary. Minimize traffic, and the probability of being jammed decreases.

There's a time to hurry up and there's a time to slow down. Know when to do which. Patience in rooting out a dug-in enemy is key. Safety lies more in hugging him closely than in being out at arms' reach. Close-in infantry using good tactics defeats a dug-in defending armor force unpro-

tected by its own infantry. Take the tanks out one at a time. Restrict the enemy's reaction with covering antiarmor fire.

Always resist the temptation to admonish Baker for his decision to go with his dismounts. Although that call may have been erroneous, the more important consideration was to preserve the independence of the battalion's subordinate commanders. The colonel knew that the majority of the decisions they would have to make would be faced by them alone, and he did not want to restrict their freedom of action and their ability to do so.

The general rule was to put the right man in charge of the right job. Franklin had been the wrong man. Everyone knew it. It didn't have to be said.

But for Always, Franklin was not the scapegoat for this costly mission. He looked to himself as the bearer of the bill. He scrutinized what he had done right and what he had done wrong. He heaped no recriminations upon himself. There was no room for that. He maintained his objectivity amidst his introspection. He needed to be in balance. By the time the review was over, the order for the next mission was ready—a night attack against the withdrawn enemy.

CHAPTER 6
Night Attack

A night attack has features all its own. Always had done many of them in his life as a light infantryman. He had thought them difficult then, although they offered many advantages to a trained soldier. He had never done one where vehicles accompanied, indeed were an integral part of, the attacking force. He was used to surprise and stealth. He was used to slow, deliberate movement over difficult terrain, terrain the enemy would have assessed as unusable to an attacking force.

He had been doing a great deal of night movement in the course of the battles of the last few days. He had moved into position to facilitate an early morning attack, into areas that were, except for the ever-present enemy reconnaissance, devoid of opposition. But this was different. Now he would have to cross the line of departure early after darkness and close with and destroy the enemy before the light of dawn. Moreover, he had very little time to plan his attack, gather his forces, and execute. It was already early into the afternoon, and night would be falling within six hours. Instinctively, he knew that he would have to use every available minute of darkness to pull off his mission. That left barely enough time even to reconstitute his devastated force. He held only slight hope that he could have them in an organized posture to push out at dark with a well-coordinated plan.

Not that he questioned the wisdom of the night attack. Always was experienced enough in the business to know that now was the time to keep up the pressure. The enemy had fought a tenacious fight to hold onto Hill 876, but with significant losses he had been forced off. He could not have had time to either reconstitute his strength or put in significant defenses by nightfall. It was doubtful that he could make any improvements after dark. It had been an exhausting fight that day, and night always has the effect of exaggerating the feelings of exhaustion brought on by a day's bitter labor. Moreover, if the enemy did attempt to improve positions at night, the noise and movement would reveal his positions to an observant attacker and would have the ultimate effect of increasing the defender's vulnerabilities.

Always' battalion, however, was badly battered from the efforts of the battles of the last several days. The activity had been nonstop, the fighting exhausting, the casualties debilitating. The infantry dismounts, at best barely sufficient when up to full strength, had been worn down to 10-man platoons. Many of the men had just come up to take the places of those lost earlier. It struck Always as a little absurd that an infantry task force could field only 60 men per company even when at 100 percent of its authorized strength. In a balanced task force of two tank and two Bradley companies, this meant that he could only hope to put 120 men on the ground. With attrition, he was now down to 60. As a company commander, he had commanded companies with two times that number of infantrymen alone. It bothered him a great deal. The machines could do only so much. Even in this age of modern, high-technology warfare the infantryman remained key. He was worth his weight in gold, yet he had been cashed in to pay for the very expensive machinery. Always needed both in his battalion. To him, the compromise did not make any sense. The prospect of a night attack was underlining the illogic of it all.

But the task force was short not only in infantrymen; it was supported by a single engineer platoon. Those men, and particularly their platoon leader, had pulled off minor miracles in the preceding days. But their work exposed them to danger. One bad moment at an obstacle under artillery fire or along a road covered with direct fire, and they could cease to exist. They numbered around thirty at top strength. A single mine-clearing mission and they could go to fifteen, or worse. Aware of this, Always had husbanded them during the defense of the day before and the attack of that morning. Tonight they would have to take their chances. There was little doubt that the withdrawing enemy had seeded mines to cover his escape routes.

If shortages of men were a great concern to the battalion commander, so was the need for replacement vehicles. Battle and maintenance casualties had taken their toll. By the end of the morning's battle they were down to less than 40 percent of their combat vehicles. Despite herculean efforts—and if the battalion was manned adequately in any single area it was in maintenance personnel—Always doubted if they could enter the night attack with more than 60 percent of their fighting vehicles. His strongest card was his antitank vehicles in E Company, which had gone relatively untouched during the last two fights. He would have to figure out a way to get maximum effectiveness from them during the night fighting. For the others, "fix forward" would be the maxim, and that meant working the people and the supply lines for all they were worth. For the hundredth time the commander felt an appreciation for his excellent executive officer, who even before the dawn attack had ended was making things happen to keep the combat power of the battalion sustained. Clearly, Major Walters was worth a combat company all by himself.

The weather was taking an unusual turn. The winds that had been rising for the last couple of days were shifting direction, and with that came a cloud cover, at first unnoticed through the swirling sand, but now darkening the sky at what would normally be the peak of the day's brilliance. Instinctively Always knew that the developments in the weather would impact on the night's operations, but it was too soon to gauge. He would have to watch and adjust his calculations accordingly.

Miraculously the majority of the key leaders were still in place. The original company commanders were still commanding, although some were debilitated by minor wounds. Two of the original executive officers were gone, and several of the platoon leaders were down. About one third of the platoons were now led by sergeants, some of these being staff sergeants, one grade below the normal level of platoon sergeant. But even though the requisite number of leaders were in place, virtually all of them were worn to the core. The activity had been nonstop, and although each man had done as much as he could to pace himself, the overall effect was not to be denied. All were beyond the point of normal physical endurance. It was will, determination, and psychological energy that was driving them now. It was getting harder for them to focus, to rationally think their way through problems. The easy was becoming difficult. The simple execution of menial tasks was now requiring intense concentration. It seemed as if sleep itself were counterproductive, since recovery from the short catnaps was taking longer and longer. Always had stressed how important it was for leaders to husband their energies, to get whatever rest was possible. He would have to remind himself to abide by the same advice.

Command Sergeant Major Hope had advised Always that night attacks here tended to be extremely tough. The vastness of the terrain, the darkness of the desert nights, the ability of the enemy to hide in every nook and cranny of his familiar territory combined to make such operations very hazardous. Nonetheless, the colonel felt confident about the upcoming operation. He had long since lost his fear of the darkness and had learned to move easily in the night. He had garnered much experience in moving light forces at night. The complication here would be the machines, but he felt he had learned much about them in the last few battles. Moreover, his determination was hardening. He figured he knew now how badly you could be hurt, and like the fighter he was, he was determined to go to his fate sobered but undaunted, intent on giving more than he got.

The objective for the night was Hill 760, some six thousand meters to the east. The distance was short enough, but the going would certainly be treacherous. The zone of attack was dominated by a wall-like ridge line to the south that pointed directly at Hill 781, the objective of Always' second battle. The cliffs were interrupted by small canyons, depressions, and rock piles that allowed for countless ambushes. The northern boundary of the zone was marked by an improved road that ran due east-west before cutting off on a line to the northwest beyond Hill 760. While this would be an easily identifiable terrain feature to help keep the elements oriented in the dark, it was likely to be mined and covered by fire. Moreover, although the road was in Always' zone, any spilling over it to the north could lead to an intermingling between him and the battalion attacking on the colonel's left. Fratricide was a major concern at night, even within the different elements of a single task force, and the danger would be tenfold should Always come close to another task force that would be as hair-triggered as his own.

It was probably this concern as much as any other that led Always toward the plan that was eventually issued to the battalion. He knew instinctively that the zone he was given was too small to maneuver his many forces safely in the dark. If they were too tightly controlled, they would be bunched up, thereby offering a lucrative target to an enemy lying in ambush. Loosely controlled, they would be prone to bumping into each other in the dark, and in the confusion opening up on one another. The answer to the dilemma lay in the narrow valley that snaked out of Hidden Valley to the southeast, the valley he had come through in his night march three days ago. Always asked for, and got, an extension of his zone to include that valley, which he immediately dubbed Route JOHN WAYNE. Intelligence believed it would be unusable, mined by the retreating enemy forces that morning. Given its narrow constriction, it was doubtful that it would be a safe route into the objective. It could be guarded by limited forces at the farthest extension of its run. Moreover, even if the valley were negotiated successfully, there remained a tortuous and confusing stretch of ground remaining between it and the final objective. The propensity for getting lost in the dark was high.

But it was the valley's very unfitness as an avenue of approach that made it appealing to Always. The enemy was likely to make the same assessment. Given the haste with which he withdrew, and his depleted forces, it was just possible that the enemy had underrated this route and either purposely left it uncovered or failed to cover it adequately. Coordination of such things was difficult even under the best of circumstances. In this case, the chances of an oversight were good.

The battalion scouts were tired. They pulled in for a quick meal, rearmed themselves, did what maintenance they could, and immediately pushed back out. Always sent half of them south to Route JOHN WAYNE; he sent the remainder 1,000 meters beyond the line of departure along the direct approach from Hill 780 to Hill 760, now broken down into two approaches, Route DIRECT NORTH and Route DIRECT SOUTH. The engineer platoon would follow the scouts moving into JOHN WAYNE, ready to clear any mines and obstacles. If they had success, then a major part of the battalion—Delta Team (one tank platoon and one Bradley platoon) and Echo Company—would follow, their objective to set up on and around Hill 781 overlooking Hill 760. Since there was only a slim chance they would be able to get through, they would constitute the supporting attack. The main attack would go either with Charlie Team along Route DIRECT NORTH or with Alpha Team along Route DIRECT SOUTH. The main effort there would be determined by the situation that developed, and would be designated by the committing of the reserve—Bravo Team (two Bradley platoons and one tank platoon)—which would follow along Route DIRECT NORTH.

On paper the forces sounded more than adequate to do the job. The reality was, however, that platoons seldom consisted of their normal complement of four vehicles. In fact vehicle shortages were now so great that the three teams leading along each route consisted only of two platoons apiece. Only the reserve team and Echo Company would be a full three-platoon force. A loss of a single platoon during the movement to the objective, therefore, would handicap the ultimate success of the mission.

Essentially, there were two ways the units could become lost—either by enemy action or by actually losing their orientation in the dark. Always would have to take his chances on the former, but he would do all he could to mitigate against the possibility of the latter. The ground surveillance radars (two were attached to the task force) were set up in such a position that they could continually vector friendly forces out to the limits of radar range, thereby sighting them in well beyond the line of departure. By being up with the scouts, they could extend their range another 1,000 meters. Both radar teams were committed in the north, where they mutually could position forces taking the direct approaches. The forces along JOHN WAYNE would have no problem identifying where they were, once in the pass. Beyond JOHN WAYNE they would have to move too quickly over rough ground for the radars to be of much good.

The scouts themselves, after conducting a zone reconnais-

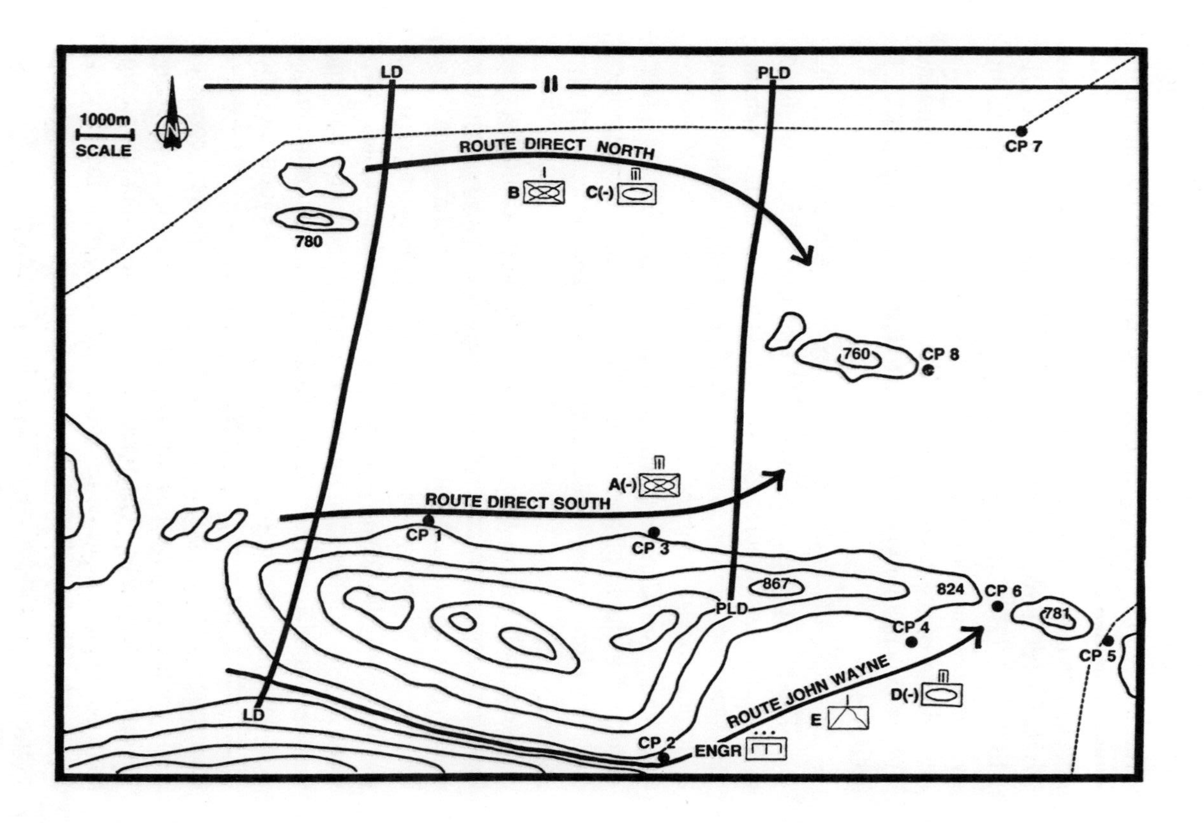
1000m
SCALE
N
LD
PLD
ROUTE DIRECT NORTH
B
C(-)
CP 7
780
760
CP 8
A(-)
ROUTE DIRECT SOUTH
CP 1
CP 3
867
824
CP 6
781
PLD
CP 4
CP 5
LD
ROUTE JOHN WAYNE
D(-)
E
CP 2
ENGR

Map 5: Night Attack

Along Route DIRECT NORTH:
Team Charlie: 1 tank platoon, 1 Bradley platoon with infantry
Team Bravo: 2 Bradley platoons with infantry, 1 tank platoon

Along Route DIRECT SOUTH:
Team Alpha: 1 Bradley platoon, 1 tank platoon

Along Route JOHN WAYNE:
Team Delta: 1 tank platoon, 1 Bradley platoon
Echo Company: 3 ITV platoons
Engineer Platoon

Note: Attrition has reduced number of platoons to two each in Teams Charlie, Alpha, and Delta.

sance out to 1,500 meters beyond the LD, would assist in keeping the bulk of the forces on target. Always did not want to push them out too far in their exhausted state. Unsupported, they would be easy prey for the enemy lurking out there. The probing for the enemy would be done by dismounted infantrymen, who in the north would walk 300 meters out in front of the lead mounted platoons. The tanks and Bradleys would be moving with their thermal sights on, able to pick up a rabbit hopping more than 2,000 meters out. They would have no problem seeing their own infantrymen, who using their compasses should be able to keep their direction in the dark. As a final safeguard, artillery spotting rounds would be fired on known locations visible from the route of march. Always' people would be able to sight these even in the dark, draw a back-azimuth from them, and calculate with a great degree of accuracy just where they were.

Even with all these precautions, however, staying on course

would still be a chore. The desert is tricky even in the daylight. At night it is downright treacherous. The cloudy skies would take away the last vestiges of light. Add to that the swirling sand, and it would be difficult to see a hand in front of your face. Leadership at the small unit level would be key.

Always had made the major decisions by about two in the afternoon. It remained for the staff to fill in the myriad of details—artillery fire support, recognition signals in the dark, passage through the line, consolidation on the objective, and so on. This was a plan that would have to be well briefed, and 1630 was set as the time for the orders group to assemble. Both Always and his S-3 had personally briefed Lieutenant Wise and his scouts, who could not wait for the orders briefing. Yet their job was so important that they had to understand exactly what was in the commander's mind.

So too were the engineers briefed. They would get to work as soon as the scouts had fixed the position of the lead obstacles and ensured that no enemy were in the immediate vicinity. At the end of the briefing the platoon leader was somewhat surprised by Always' admonishment to preserve his force. He could not know that his commander was already looking ahead to the next mission in which the engineers would be even more critical.

And so, even as the orders group meeting started, the scouts and engineers were hard at work breaking through the first obstacle along JOHN WAYNE pass. It was noteworthy that no enemy was spotted in the area.

The TOC, which had moved up to a wadi just short of the line of departure, gave off an atrocious odor. The gathering of staff and commanders, long without a shower or change of clothes, exuded an aroma so foul it would turn back a rat in midstride. They were spared the repulsiveness of their own stink only by their long-conditioned acceptance of such an extreme state. Immunized by their own stench, they could tolerate the noxiousness of the group. A clean soldier would have been nauseated. Moreover, his very cleanness would have made him an outcast from the group. Despite the rigorous discipline that demanded a shave and a washing every day, the battalion had grown beyond the pale of civilization. Their filth was imbedded in their pores. Their clothes, soaked by sweat, bleach dried in the heat, and soaked by sweat again, were uncleanable. Their soiled condition seemed to go with the red-eyed, fatigue-blackened visages of the assembled group, and was worn as if a badge of honor. Amidst this putridness, the order was briefed.

"Expect a counterattack." Always was glaring at his commanders. "They don't want to give any more ground. When we take it from them, they'll want it back." He looked for recognition of his words.

It was getting harder for his people to look ahead. The task at hand was always of such prime importance, always such a matter of life and death. It was hard for company commanders, better yet platoon leaders, to be thinking beyond the current mission even when they were fresh. Now that fatigue had settled over them like a poisonous cloud, the here and now was taking even greater amounts of concentration.

"Sir, what do we when we spot the enemy?" It was Baker.

"You kill him." Always was slightly annoyed.

"What about surprise? Don't you want to keep the surprise?" It was Archer.

"I figure he knows we're coming. He just doesn't know where and how. He'll probably see us before we see him, and that will generate an immediate report. Kill them as you come to them and that will leave him fewer to defend with. Besides, if he sees us in the north and focuses there, then we have a better chance of flanking him from JOHN WAYNE. Any way I slice it, it makes better sense to kill them as you come to them."

"What do you think he'll counterattack with, sir?" It was Evans, the freshest of the commanders.

"I don't know. Can't be sure. Probably anything he's got handy. Could be a company. Maybe a battalion." The commander was pleased with the question, glad at least one man was alert enough to think of the future.

Future. Future. Only in war could the future be considered 0400 the next morning. It underlined how fleeting life was. You lived by the hour, indeed, by the minute. If you lived to nightfall, then you could plan ahead to morning. And tomorrow? Well, that was a whole 'nother day—a lifetime!

Always snapped himself back from his reverie. God, he was tired. "Have to stop musing," he thought to himself. "Have to focus on the mission."

"Captain Dilger, Captain Evans, I want you to hold your positions where they are now. I may not be able to get you through the pass. If not, then I'll bring you up along the direct routes. No need for you to bunch up behind the scouts. If they get through, then I'll decide if there's time to move you by way of the pass." The colonel was trying to avoid sounding doubtful, but he had to hedge his bets. The gamble was bold enough as it was; no need to compound it.

"Look men, what we've got here is a plan. I think for the most part it will work. But a thousand things are going to happen out there between now and tomorrow morning. It's going to be confusing. You can bet that we'll be knocked off the radios for at least part of the time. You've all got to understand where we're going and what to do when we get there. Commanders, as always, I expect you to take command. Push this attack through. Don't let it get bogged down. I can assure you I'm pressing on. I plan to kill any enemy I find in my way, and I plan to be there in the morning. If you all end up where I said I wanted you, all well and good. If you don't, I plan to be there anyway. So get there and take part in the fight. There will be enough enemy to go around."

By 1730 the meeting ended. The attack was set for 2000,

giving the task force eight hours of darkness. It would take that long to move everybody into final position. A light rain was falling on the TOC canvas as the smelly group broke up.

Always found the time to visit his companies. On the surface they seemed in a shambles: replacements coming up hurriedly; vehicles being towed and worked on everywhere; ammunition being loaded by bone-weary men who looked as if they might drop from the effort; sergeants growling at men to move here and there; lieutenants draped over worn map sheets, giving instructions to squad leaders and planning fires with their artillery forward observers; smelly, greasy fuelers rushing about to pour diesel into thirsty machines; aid stations littered with bloody bandages, processing casualties; combat veterans snatching a moment's sleep wherever they could safely get out of the way of roaring vehicles. But as he talked and looked, Always heard and saw the deeper signs that revealed a battle-hardened unit ready for more combat. Weapons were clean; soldiers were fed; vehicle load plans were adjusted; the orders were passed; information was shared; and everywhere there was a solid determination to do the mission, to defeat the enemy.

The going was slow in the pass. One obstacle had been cleared, only to reveal another, then yet another. It would be hours and hours of work at best. They might never get through. Always decided to stick to the plan. He could adjust it later if need be.

By 1930 the commander had moved to a position just short of the LD. The rain had picked up—hardly a drenching rain but enough to make itself felt and to make a man want to hunker down. That was good, Always figured. The defenders would think about it more than the attackers.

A net call went out at 1945; intelligence was updated and slight adjustments to the order were given. By 1955 all parties were off the net. The self-imposed silence would cause no immediate problems. The first 1,000 meters were to be done in radio silence, the movement controlled by direct command and by wire being played out now by the advancing infantry. Nonetheless, the frequency was shifted to the alternate. It would serve as a backup until the full extent of the wire was played out, at which time the radio would again become the primary means of communication.

By 2010 Always' track lurched into motion behind Team Charlie, which along with Alpha was already ten minutes into the attack. Close behind him came his artillery and air force officers, and behind them at about 200 meters came Bravo.

For the first hour things went quietly. Progress was good, if not steady, vehicles having to stop and wait for the infantry to keep pushing out in front of them, the infantry moving slowly and with care. That was the way Always wanted it. There was no rush to get there.

Sometime after 2100 the first contact with the enemy was made by Team Alpha. An enemy BRDM moving from west to east in the vicinity of Checkpoint 1 had appeared in the thermal sights of an Alpha Bradley. It was moving with no great dispatch, probably lost in the dark looking for one of its unit's positions. Captain Archer had been informed and was scanning with his own thermals to make sure it was in fact an enemy vehicle. The silhouette is distinct to the trained eye, but the tired and the new—and the gunner's eyes were certainly one or the other—could make a mistake. Satisfied that his men had enemy in sight, Archer gave the okay to open fire. Six 25mm rounds spit out of the designated Bradley and the enemy vehicle was destroyed, its crew dead. Alpha received some sporadic and poorly adjusted artillery fire for its efforts, none of which did any damage.

Hasty mine fields were being discovered along both direct routes. Because of the flatness of the terrain, the mine fields were not tied in to any restrictive terrain features, and so, with patience, were able to be bypassed. This cost time, however, and soldiers as well, since guides from the ranks of the infantrymen would have to be left in place to ensure that following vehicles did not blunder into the mines. In the meantime, no word had been heard from the scouts and engineers in the pass for more than two hours. The terrain was just too steep and broken to allow for an uninterrupted radio transmission.

At the time of the Alpha contact the radios on the battalion net had gone back into operation. The result was another jam session, Always fighting his way through it for a while before passing the order to switch to the second alternate. His head had begun to hurt from the racket in his ears.

By 2230 the pattern for the night had established itself: sporadic fighting followed by momentary confusion as commanders sorted out what was happening. Infantrymen in the lead proceeded with caution; drivers in vehicles fell asleep at their posts as they waited for the orders to move on. Always had left his vehicle twice to prod the trail platoon of Charlie. Both times he had found drivers and vehicle commanders dozing at their instruments. Once he had run the two hundred meters back to Bravo, not wanting to risk approaching his own men in his vehicle from the opposite direction, woke the lead vehicle crew members, and got them moving apace with Charlie's progress. Again the enemy intercepted his radio net and jammed it. For the third time the battalion shifted frequency. And so it went. Stop, start, a brief firefight, a stall, get people and vehicles moving, get jammed, work through it, shift frequency, run to the unit ahead, run to the unit behind, keep everybody moving forward.

Always was consuming energy at an alarming rate. His headache had developed into a constant throbbing, his neck and back ached, his face and hands were chapped and bleeding. Each time he climbed from his vehicle his knees protested. Each time he ran in the sand his legs rebelled. He was driving himself now with all the willpower he could muster. He found himself cursing aloud at the enemy jamming his radio and tried to settle

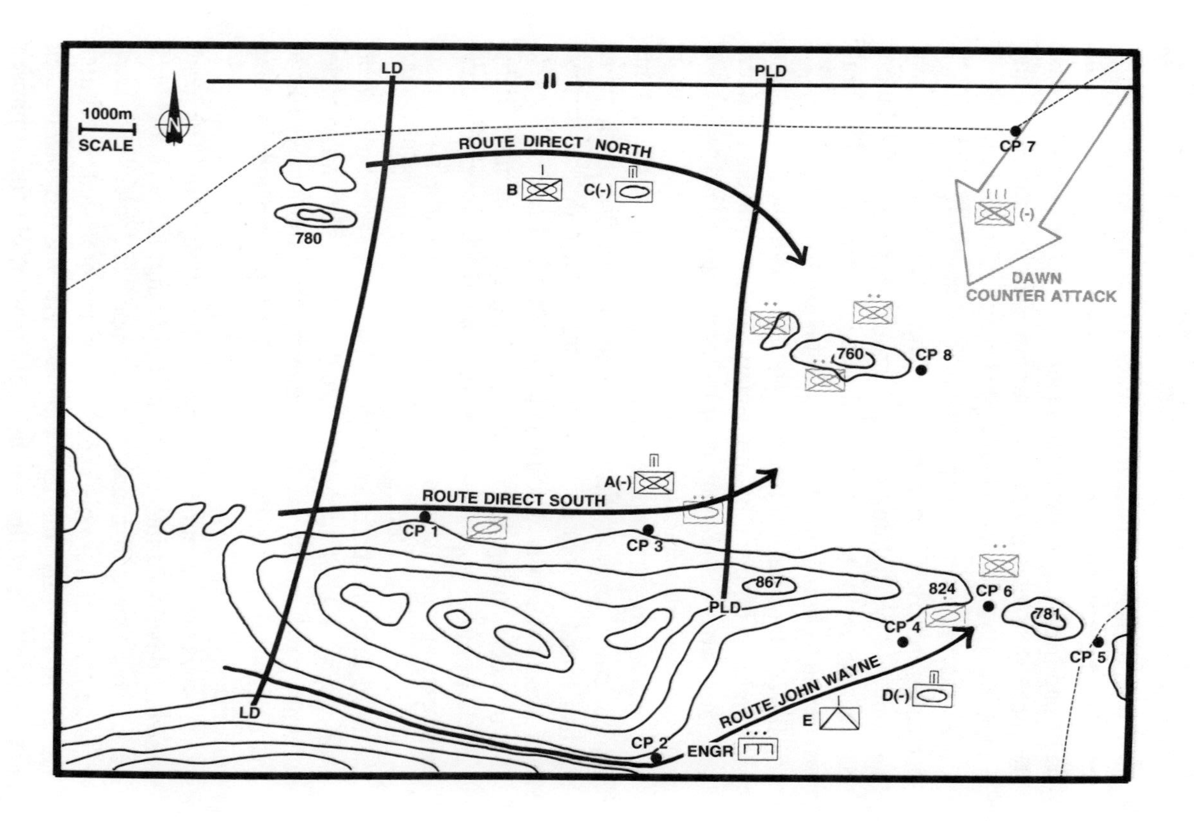
1000m
SCALE
LD
PLD
ROUTE DIRECT NORTH
B
C(-)
780
CP 7
(-)
DAWN
COUNTER ATTACK
760
CP 8
A(-)
ROUTE DIRECT SOUTH
CP 1
CP 3
867
PLD
824
CP 6
781
CP 4
CP 5
ROUTE JOHN WAYNE
D(-)
E
LD
CP 2
ENGR

himself in order to conserve strength. It was a losing battle as he caught himself shouting at the sleeping crews in their tanks, banging on their armored vehicles with his helmet. His chemical suit, drenched by rain water each time he left his Bradley, became heavy and sodden, and his clothes underneath were drenched in sweat. From 2345 to 2400 he talked through a screeching, high-pitched jamming session, hoping to hold out on the established net until the standard frequency change at midnight, only to discover that the enemy was only a few seconds behind him in arriving at the new day's net. His frustrations were mounting.

But they had made progress. Only two vehicles and a few infantrymen had been lost to enemy fire in the north. As far as he could tell, all remaining vehicles and soldiers were still en route, headed in the right direction. It was from the units in JOHN WAYNE pass that he had heard nothing. By midnight this was a critical vacuum, its silence deafening. Accordingly, Always ordered Captain Dilger to send out his lead platoon leader to make contact with the scouts. Soon he would have to decide which way to commit his forces. He wished his headache would go away.

Always could not know the consternation he was causing the enemy, who had expected to be attacked from along the road. When he came under fire from Alpha Team along Route DIRECT SOUTH, it threw him into disarray. His artillery had not been planned there. Most of his night vision devices had been situated in the north. The remainder had been left behind to watch the pass. The continued pressure by Always' battalion caused him to shift his forces from Hill 781 to Hill 760. The order had been passed so that the night vision devices could be moved to where the action was breaking. In the confusion that dark often brings, the order had precipitated a withdrawal of the enemy from his position around Checkpoint 2. By the time the enemy commander discovered the error, he concluded it was all for the better. Repeated sightings of enemy infantrymen and vehicles had inflated the reports, convincing him that the entire effort was being made along the direct approach. He was quickly shifting his forces south and west of Hill 760 to stop Always' attack.

When Delta's Lieutenant Sampson made contact with the engineers clearing the last obstacle at CP 2, the enemy was long gone, a fact confirmed by the scouts. He could get no assurance that yet more obstacles, perhaps covered by enemy, did not lie between him and Hill 781. Momentarily he was thrown into a quandary.

His orders had been to link up with the scouts to determine if the way was clear. He understood that the decision to commit Delta and Echo rested with that information. He also understood that time was fleeting, that if he waited for absolute confirmation, the moment would have passed. In order for the two companies to get up to Hill 781 in time to assist the assault on 760, he would have to speed back and bring them up now. Yet if he did that and the way was unclear, they would never be in position in time to help the action. Worse yet, they could run into a buzz saw and get chopped to pieces.

Sampson considered the alternatives. He was smart enough to understand that there was a convenient way to avoid the responsibility. He need only to follow out his orders to the letter, which meant waiting until he could get a clear picture. He certainly could not be faulted for that.

But escapism was not in him. He felt a responsibility for this action. This feeling of obligation had been put there by his company commander. It had been put there by the few minutes he had talked with Lieutenant Colonel Always. It had been put there by his training. Most of all, it had been put there by his devotion to his men. He made a decision. If the enemy had not taken advantage of the best position from which to defend, then most likely he would not be on the poor ground between

CP 2 and Hill 781. Even if the enemy had put obstacles out there, the ground was open enough to find a way around them. It was now more a problem of terrain navigation than anything else. And he was confident that he could find his way to the objective. He decided to go back, give his opinion that the way was open, and recommend that Delta and Echo come down JOHN WAYNE pass. In that decision, Lieutenant Sampson gave the battalion the chance it needed to take 760.

When Always got word that the pass was open, it was already 0100. He studied his map and tried to calculate the time-distance factors. His mind was finding it difficult to focus. Twice he lost track of what he was doing. On the third try, and only with great effort, he was able to focus long enough to determine that without resistance and without getting lost, they could close on Hill 781 by 0330. He gave the order for Delta and Echo to move out, and immediately got jammed. He never even heard if they rogered his order. By the time he had shifted frequency again, they were gone.

Charlie and Alpha came abreast of the probable line of deployment (PLD) at 0234. There had been one heavy fight in the vicinity of Checkpoint 3, where a platoon of tanks had established an ambush position. But the opposing tanks had opened fire early and nervously, and Captain Archer had been able to fix their positions by their fire and move his infantrymen against them. In such a face-off, if the infantrymen can stay under cover until they close, and if there is no enemy infantry to thicken the defense, the advantage lies with the attacker. Accordingly, the enemy had stood his ground until the loss of the first tank told him that he was compromised. With that he had pulled out his remaining three tanks, two of which were destroyed immediately by the Abrams tanks attached to Captain Archer.

Always had correctly deduced that the bulk of the defending enemy was to the south of Hill 760. In his mind this dictated that he should commit Bravo in the north, behind Charlie, the principle being strength against weakness. Alpha would have to conduct a fixing attack. It would be rough going for Captain Archer, but he was a solid commander and the best bet for such a high-risk operation. If the main attack went in violently enough, it would relieve the pressure on Archer before doing him irreparable damage. If only he knew what Evans and Dilger were up to! He had sent Major Rogers after them when he could not gain radio contact, but so far all that had gotten him was the loss of Rogers.

But this was no time for equivocation. The infantrymen would begin to make their dismounted approach immediately. At 0310 the combat vehicles would cross the PLD. Carter would press on to CP 8 while Baker went into the north side of Hill 760. Archer would assault the south side of 760, then move on to link up with Carter, Bravo sweeping the objective to make sure all enemy were dead or captured. Infantry forces would give their exact location at 0305 so that an artillery preparation could soften the defenses on 760.

The last five minutes of the order had been passed through a barrage of jamming, bleeps, whistles, stuttering, and screeches. Always dismounted his Bradley once again to find Carter and Baker in their vehicles, to ensure that they, at least, had gotten the order. He was determined to make good on his vow. He would be on the objective by first light, due at a few minutes after four. If he could have at least Bravo and Charlie with him, so much the better.

A reinforced enemy motorized rifle company had begun the defense of 760. The loss of a platoon of tanks had reduced that to approximately a company-sized element, two platoons of which were now out of position on the south side of the hill. A few reconnaissance elements in the area had added their support to the defenders, one of them destroyed while trying to link up with an outpost at CP 3. The BRDM at CP 4 was the only element guarding the backside approach. When the artillery started falling at 0325 he shifted to CP 6, to see better what was happening atop 760. He was afraid he might be abandoned at his post in the rush of the night fighting. His commander's voice was sounding increasingly shaky as the night progressed.

The task force's infantry had found and breached the thin wire obstacles on both sides of Hill 760. According to the prearranged signal, they marked them with chemical lights. It was through this breach that Charlie and Alpha rushed in at 0330. Instantly the night was ablaze with flashes of light. The two enemy platoons in the south put up a stiff resistance for the first few moments. But their positions were unprepared, having been shifted during the confusion of the night. They did not have the advantage they otherwise would have. Two of Alpha's Bradleys were hit, one of them Archer's, killing the Stinger gunner in the rear and leaving the crew uninjured. But the enemy took two hits as well. Their platoon leaders, tough veterans of many battles, remained unshaken. The position there hung in the balance.

At that moment the reconnaissance vehicle at CP 6 sighted Delta making a swift approach from the south, Lieutenant Sampson in the lead. The enemy scout yelled a warning over his company net, then pulled out toward 760. One of the defender's BMPs saw the movement and, not believing there to be any of his compatriots in that direction, opened fire on him. The BRDM, now in a state of panic, pursued by two companies of Always' battalion and drawing fire from his own company, returned fire at the BMP, nearly hitting it and sending shrapnel pinging off other defenders. That was the straw that broke the back of the defense. Caught in a cross fire and alarmed by the report of a large enemy force approaching unimpeded from the south, the defenders pulled out to the northwest. The movement brought them by the rear of their own platoon to the north of 760 and across the front of Charlie Company coming in at CP 8. The startling effect unhinged the northern platoon, which joined in the dash for safety. By 0336 Hill 760 was taken. Always had his prize, and at little cost.

He took the reports with a sense of elation. Bravo was sweeping over the top of the hill. The infantrymen were coming up to link up with their vehicles. Delta and Echo were consolidating on 781. The engineers and the scouts were fast closing in on them.

It was in the next few minutes that fatigue overtook the battalion. The radios had gone silent, all necessary reports having been rendered. The infantry went to ground, waiting for the final few minutes of darkness to pass before they found their vehicles. The vehicle crews paused for the infantrymen and the rest of their units to close together. They became frozen in a world halfway between light and dark, excusing their inactivity with the implication that each was to wait for the other. One by one, men dropped off to sleep. Leaders, allowing themselves a moment's pause after the exertions of the night attack, at first did not notice the stillness, then succumbed to fatigue in varying degrees themselves. Always' warning of a counterattack was a forgotten statement.

The colonel too let down his guard. Climbing down from his vehicle command post, he made a call to Major Walters on the administration and logistics net from the radio in the back of the Bradley. The driver had dropped the ramp, and the vehicle was darkened to maintain light security. Always encouraged Major Walters to bring up whatever vehicles had been repaired during the night to Hill 760 just before dawn, and his XO responded that he would comply. The commander looked at his watch, saw it was 0346, and allowed himself a short moment to catch his breath. In the stillness of the early morning darkness, lulled by the whirring of the Bradley engine, he slipped into slumber.

A last spate of will, his subconscious mind a residue of his conscious determination, jerked him to sudden wakefulness. He glanced at his watch. It was 0410. A faint light was coming into the sky. "Damn! The counterattack!" Always climbed back into the cupola.

For five minutes he called frantically on the radio. No commander answered. Despite his best efforts he could get only the TOC, Major Walters, and the air defense platoon leader. He ordered all three to keep trying by whatever means possible to raise the unit, then he ran to the artillery officer's personnel carrier. He climbed the track and banged harshly on the hatch. "Wake up! Wake up! God damn it, wake up!" The sky was showing a faint blackish blue. It was getting lighter.

The air force officer stuck out his head. Always yelled at him. "Get everybody up in there, then start moving around waking people up. Tell them to get ready."

He received a dumbfounded look in return. "Damn it! Do what I tell you. Get moving!" Always wanted to hit him. The officer started scrambling to get the others up.

Always ran back to his own track, grabbed Spivey by the arm, and shook him awake. "Come on. Let's go. We've got to get the battalion up." On the way into his cupola he poked Sergeant Kelso, also asleep at his position on the gun. The anger, if not his determination to get the battalion ready to fight, was now leaving Always.

He had made it to five positions when the first signs of dust appeared several kilometers off to the northwest. The horizon was now light enough so that the dust trails were visible. The enemy was coming, and there were a lot of them.

A few of the leaders were up and about now, struggling desperately to wake the others. It was as if the task force had been hit with a potent sleeping gas. Soldiers were knocked out everywhere. Now, Bradleys were pulling up beside other Bradleys, firing their 25mm guns at the sky, trying to save a precious few seconds in waking up sleeping crews. Slowly the radio nets came to life. Hill 760 would be awake for the fight, but it would not be consolidated. It would be a free-for-all. The few minutes before and after 0400 would have made all the difference in the world. What would have been a coordinated defense against a desperate counterattack now became a meeting engagement at best. Worst of all, the forces on 781 were too far away to do any good until 760 had been overrun. Always stared at the oncoming forces. He estimated about two battalions were racing to close with him.

At last Major Rogers came up on the radio. Always told him to take his forces on 781 and bring them into the melee around 760. Rogers said it would take him about ten minutes to get everybody ready. Always cursed again, then turned to make his fight.

It was a donnybrook. The enemy came in line formation, his artillery crashing in on Hill 760. Always countered with his artillery on their moving formations. The enemy's was more effective. The defending enemy company commander had sent in his fire plan to his commander earlier in the evening. It had included targets in and around Hill 760. This planning now eased the enemy's calls for fire. Always had to improvise, making use of the unfleshed-out details of the counterattack plan.

The direct fire engagement went more toward Always, his vehicles stationary, able to pick out targets appearing amidst the dust and smoke. The enemy was unable to fire on the move. He needed to get set to send home his rounds. That was enough to cost him the first two companies.

But there was more enemy coming on. Perhaps seventy to eighty vehicles were closing at top speed. Always was starting to take losses. Rogers got in motion and made a dash toward CP 8. For perhaps seven minutes the battle raged back and forth. Always, seeing rounds careen all around him, hearing the roar of the enemy tanks, knew that he could not hold. The enemy had the ground if he wanted it. If Always stayed, his task force would be decimated. He would have to pull back, give up his hard-earned objective. The thought was devastating.

It was at that moment that the enemy pulled back. He knew he could take 760, knew he had Always. But there had been another battalion attacking with Always. It had been unsuccessful during his night attack, in fact had gotten lost in the dark, but with the light it might reappear and shift the odds. If that battalion broke through, then the lines of communication of the counterattacking force would be threatened. As it was, Always had put up just enough of a defense to make holding 760 untenable for the enemy. Bitterly, the commander of the counterattacking force passed the order to fall back. He would have to wait for another day.

The after-action review at 0900 was hard. But not because the observers were harsh. Lieutenant Colonel Drivon was surprisingly gentle, in fact heaped much praise on Lieutenant Colonel Always for the aggressiveness of his night attack. So much had gone right, only the consolidation on the objective had gone astray. But it was enough.

Always was disgusted with himself. He had had his victory, but through his own weakness he had let it slip away. Or had it been weakness? Perhaps he had taken too much upon himself. Perhaps he had trained the battalion to respond to his authority so much that when he let down for an instant his men took it as a signal to let down as well. Maybe it was his very strength that had turned into a weakness. He had pushed himself relentlessly. What little energy he had left he had dissipated in the exertions of the night attack, the struggle with the jamming on the radio, the sprints back and forth to keep the companies moving, the endless vigilance to ensure that the battalion was doing all that it should. Perhaps it was in that effort that he had set up the disaster that took place between 0346 and 0430, when he finally got enough of the battalion back on its feet to make a fight, albeit an inadequate one.

Certainly there was enough strong leadership in the battalion to allow others to assert themselves. Walters could lead the battalion in a heartbeat. Rogers as well. The commanders were all solid, ready to assert themselves. Even the lieutenants were ready to lead. Hadn't Sampson made a brilliant and courageous decision during the night? Hadn't Lieutenant Wise pulled off coup after coup? The battalion was full of good men. Always had given them leadership. Perhaps it was time to share in that leadership, to let them take the reins a bit.

The lesson to be taken from this was the hardest one of all:

A commander is human, and as a human, he is limited. He cannot shoulder the entire burden by himself. He needs others to help him, to pick up where his energies run out, when he cannot be there, when he is hit. He can command—but he needs others to make his command effective. It is not a sign of weakness to let others assert their strengths. The weakness lies in excusing them from the responsibility of independent, decentralized leadership, from denying them the incentive to pick up the mantle when the commander cannot do it all himself.

It was a startling idea to Always. His entire life he had shouldered his responsibilities himself. He jealously guarded them, as if they and his ability to tote them around were a measure of his manhood. To understand that the greater wisdom, the greater measure of his worth, lay in his ability to engage others in their carrying was like the lifting of a great veil from his mind. Of all the after-action review lessons, perhaps this was the greatest one of all. Always took it to heart.

CHAPTER 7

Battle Position Defense

The mission was to defend Hill 781 and Hill 760 from a battle position. This was different from the previous defense in that a battle position defense is not as fixed as a defense in sector, not necessarily tied in with friendly units on the left and right, and affords greater flexibility to the commander. The objective is the same, however—to allow no enemy to pass and to kill as many of them as possible in the process.

Always' immediate dilemma was that the enemy could take either of two directions. The first would take him to the east of Hill 760 right at Hill 781 and on to the south. This would be the more serious combat, as the terrain constricted on either side of 781 in such a way that a fierce fight would be forced as the enemy tried to penetrate. The second would take the enemy past Hill 760 and on to the west, a route that could avoid the bulk of Always' battalion unless he chose to place them along that avenue of approach, the least defensible terrain.

Always did not want to fight the enemy forward (north) of Hill 760. The ground was too open, and a massed motorized rifle regiment, which was what intelligence told him to expect, could smash through him on the open ground. Yet if he pulled all of his forces back to 781, that avenue of approach would be a free ride for an enemy staying out of range. The friendly battalion that had attacked on Always' left during the night attack had not made it to its objective and was now held up in the vicinity of Hill 876. Always could let it take the brunt of the attack passing by him, but in so doing he would have avoided his mission, the battle position defense of both 760 and 781. Accordingly, such a cowardly course of action was out of the question.

On the other hand, Always wanted the fight to break in and around Hill 781. That little hillock, splitting the passes at CP 6 and CP 5, offered a major attraction for a defensive action. The defenders would have to do a thorough reconnaissance of the area, but early indications were that it was ideal terrain for a cunning defense.

It was Captain Johnson, the assistant S-3, who first proposed the possible solution to the dilemma. "Sir, if we could put a substantial force forward at the base of 760, say on the north and east where the enemy would be sure to see it, we might lure him into attacking through them into 781."

"What are you getting at, Captain Johnson?" Always wanted him to elaborate on his idea.

"Well, sir, if we put a force up there it denies the enemy any chance of bypassing to the north on a beeline to 876 in the east. A company or two positioned there could rip any movement like that to shreds from flanking positions, and the survivors would only have to fight our sister battalion in the end anyway. But if he saw a force up there, on ground that is not too good for defense, then he would probably figure he could smash right through if he came head-on at it. After that he wouldn't expect much left around Hill 781 to hold him up."

"You've got a point there, Captain Johnson, except that the enemy calculations would be correct. If I put two companies up on 760 then I couldn't hold at 781 if that's where the main battle broke." Always was probing, hoping Johnson had thought it through a bit more.

He had. "But sir, I didn't say it had to be two companies. And maybe we wouldn't have to leave them up there once the action started."

"You mean we could withdraw them in time to thicken our defenses around 781?"

"Yes, sir. We would have to wait until the enemy was committed, you know, deployed on line and attacking so that it would be almost impossible for him to pull out of it and change direction. Then we could pull back to a prepared position and let the wave break on our defense in and around 781." The young captain's eyes were bright, gleaming, staring intently at Always.

The tired lieutenant colonel was mulling over the rapid-fire ideas in his head. A spark had begun to glow in his heart, and it was catching fire in his mind. The intensity of the captain was striking. He was onto something there. He was advocating risk, big risk, and he was not letting Always carry it all himself. He was willing to acknowledge authorship for the idea, to shoulder part of the responsibility if the plan did not work. He wanted to contribute, to help his commander see the options as he saw them. He wanted to beat the enemy, to fool him, to draw him in, to pound him to pieces on the best ground available. He was willing to go on record stating just how to do that.

Always' mind was leaping; he knew that the captain was right. And he could see beyond the initial move, into the placement of the companies, the preparation of the obstacles, the size of the force to put on 760, when to withdraw it, where to put it when it withdrew, how to further encourage the enemy to keep on toward 781 by flanking him with artillery fire that would discourage any deviation from his path, and where and when to strike with the reserves. But before he spoke he remembered his lesson of the previous battles. How much better to get subordinates to share in the decision making, the problem solving, the responsibility bearing. The youth of the captain alone indicated that he could bring much greater energy to the solution of the problem than could the battered lieutenant colonel. And not just him. The rest of them—the captains, lieutenants, sergeants, and soldiers—were young men, as dedicated and committed as Always himself. Given the freedom to share in the decisions on how best to fight the battle, they would probably spare no energy in exploring every option. Always could correct any missteps, any misdirected exuberance. The important thing was to get them all in on the ground floor, to let them use their minds, their bodies, their skills and their energies in determining how best to fight the enemy. If it was their idea, not only might it be good, it might commit them that much more to its execution.

"Captain Johnson, I think you've got something there. Let's get the staff together on top of Hill 781 and you can discuss your idea with them. Thirty minutes after that we'll pull in the company commanders and give them some preliminary orders for their placement just so they can orient and position their forces. We've got until the morning after tomorrow to complete our defenses. After they've looked the ground over this afternoon we'll meet again tonight and listen to their ideas on fleshing out your plan. The only early recommendations I'll want from the staff is where to put in the major obstacles, and that so I can do some early work by our engineers and whatever attached bulldozers you can scrape up for us."

"Yes, sir. What time do you want to give the order?"

"Please get with the S-3 and the XO on that and give me a recommendation. I don't want to lose the opportunity to work tonight, but I also want to give everybody a chance to look over the problem. I need some help in figuring this one out."

"Yes, sir." Johnson's voice belied his enthusiasm and pride.

The preliminary meetings were over by 1400. The commanders and staff suggested an initial meeting for the operations order at 1800, with a second meeting, more or less an update, to take place sometime in the afternoon of the following day. Always accepted the recommendation and refrained from giving orders except to assign D Company the mission of defending forward (the actual composition of the company to be decided at the operations order) and to direct two battalion obstacles to go in on either flank of 781. Captain Johnson would directly supervise the location, in coordination with the engineer platoon leader.

As the battalion leaders took off in all directions to conduct their reconnaissance and pass the orders to bring their units up to the designated company battle positions (general areas suggested by the assembled group and approved by Always with minor modifications), the battalion commander moved to his jeep to take a forty-five-minute nap. He would get up in time to take a short reconnaissance of the area prior to the operations order meeting at 1800. His subordinates, he was sure, would bring him the details of the terrain he needed to know but had not necessarily seen for himself.

His men did their work well. By the orders meeting they had covered their respective areas well and were not at all reluctant to offer suggestions as to how to put in the defense. Captain Johnson agreed that either side of Hill 781 was a good place for emplacing obstacles, but felt that a gap would have to be left to allow Delta to withdraw. The S-2 suggested leaving the gap completely unprepared, thereby drawing the enemy toward it in an attempt to follow Delta. The artillery officer offered the view that he could close the gap with FASCAM (artillery-delivered mines), not only denying the enemy passage but catching him midstride in the mine field.

Captain Dilger declared that he could ensure forcing the enemy to deploy and still cover his withdrawal if he were given a four-platoon force—two tank, one Bradley, and one antitank. He explained how he would deploy them in depth to allow with-

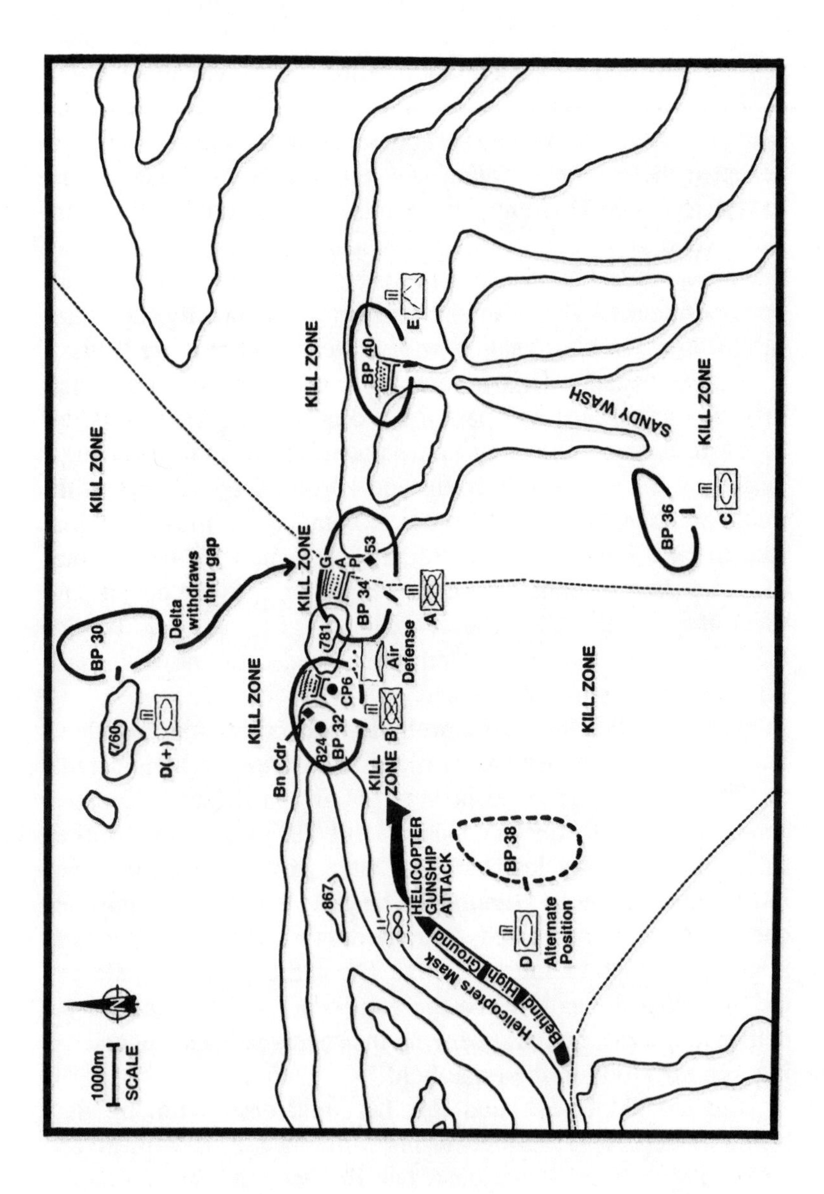
1000m
SCALE
N
Helicopters Mask Behind High Ground
HELICOPTER GUNSHIP ATTACK
867
D
Alternate Position
BP 38
KILL ZONE
KILL ZONE
Bn Cdr
824
BP 32
CP6
B
Air Defense
781
BP 34
A
G
A
P
53
KILL ZONE
KILL ZONE
760
D(+)
BP 30
Delta withdraws thru gap
KILL ZONE
KILL ZONE
BP 40
E
SANDY WASH
BP 36
C
KILL ZONE

Map 6. Last Battle

Team Delta, reinforced, forces enemy to deploy for attack early, deceives him as to main line of defense, then withdraws through gap at edge of obstacle, dropping off a Bradley platoon to thicken Battle Position 34. Delta then defends in depth from BP 38.

Major Rogers closes gap with artillery-delivered mines after Team Delta comes through. At this time, all routes to the south are closed and covered by fire.

Major Walters fights the deep battle, shifting Delta and Charlie as necessary, should the enemy penetrate.

Lt. Col. Always is positioned on the left flank of Battle Position 32 where he can see entire battle unfold.

Helicopter Gunship Battalion approaches from west, masked behind high ground; adds fires to kill zones north and south of BP 32 as needed.

Team Echo closes route into Sandy Wash from BP 40. Obstacle belt covered by fire, defended in depth.

Team Alpha defends from Battle Position 34, gives no ground east of Hill 781.

Air Defense ambushes enemy air from behind high ground, vicinity Hill 781 (BP 74).

Scouts deploy well forward, give early warning to Team Delta.

Engineers, with bulldozers attached, dig in defense (vehicles) and emplace battalion directed obstacles.

Heavy artillery concentrations along enemy avenues of approach are planned by battalion and assigned to specific companies/teams to be called.

drawal by bounds, and what signals it would take to get them moving in time. The artillery officer had worked up a plan to slow the enemy's attack with fire and cover Dilger's withdrawal by smoke.

Captain Baker, positioned in BP 32, argued for control over the development of the battalion obstacle by CP 6. It was in his battle position, and he could best integrate the fires covering it. He would use it as a linchpin in his own defensive scheme.

Similarly, Captain Archer argued for control over the obstacle in his battle position (BP 34). For a while a debate raged over who had the responsibility to close the gap, Archer wanting to do it since it was in his area, Dilger wanting to do it since he was withdrawing through it, and the staff wanting to control it since it was key to the battalion's defense. Always directed that Archer could coordinate its emplacement but that Dilger and Major Rogers would jointly agree when to call for the FASCAM (the artillery-delivered mines) from a position adjacent to the gap. Further arrangements were made should one or both of them become casualties.

Both Archer and Evans had discovered a hitherto unnoticed avenue of approach to the east of BP 34. A narrow canyon broke out to the south on the east flank of the battalion, and would have to be watched by antitank weapons and infantrymen. Since Archer would be busy enough at BP 34, commanders and staff agreed that Evans could best cover this sandy wash area if he were given an attached platoon of infantrymen, who would come from A Company minus their Bradleys. They and the ITVs (antitank tracks) would cover that approach. Additionally, the mortar platoon would be in direct support of Captain Evans should the enemy try to dismount and work his way through the canyon.

Charlie Company would defend from BP 36, covering any penetration down the sandy wash through BP 40 or through the gap at BP 34. Carter felt that both these were bets that should be hedged, and suggested he could do the job with a tank platoon, a Bradley platoon, and one platoon of ITVs. This gave him the range to cover any avenue of approach breaking out to the south. Always agreed and gave him a platoon of Bradleys from Captain Archer, the infantrymen from that unit already having gone to Echo.

The air defense platoon leader offered that his best forward defense against enemy air was up on Hill 760, but was concerned that he would be stuck out there when Delta withdrew. After some discussion by the S-2 and the S-3, it was determined that initially a section of Vulcans (two-tracked vehicles) would deploy forward, adding to the deception of a forward defense, but under cover of darkness on the second night would fall back to the reverse slope of 781. Here they would not be seen by the attacking enemy, would have some cover from an air attack, and would have a chance at ambushing the aircraft as they passed over the main battle area from north to south. The other section of Vulcans would track and fire from BP 34, while the Stinger missiles would cover each of the maneuver companies. It was somewhat risky, but if it worked it would surprise the enemy air forces.

The aviation battalion commander was in attendance at the meeting and held a discussion with the commanders. Major Walters joined in, as did the air force liaison officer. Agreement was reached that helicopters approaching from the west, staying low and to the south of Hill 867, could remain unseen until they were able to bring their fires to bear in the passes on either side of Hill 781. Moreover, should the enemy penetrate, they could quickly cut south and chew them up from behind. The helicopter commander promised to return the following day with his staff and company commanders to flesh out the plan and, upon Always' request, to conduct a rehearsal of his planned approaches, mixed in with a few phony rehearsals to delude the watching enemy.

Always listened attentively, somewhat refreshed by his short nap, and chimed in when a decision needed to be made or when he felt it necessary to redirect the discussions at hand. The meeting took a little longer than normal, and when it broke up, Always held a smaller meeting with his commanders and his S-2. They reviewed some of the bidding again and closely coordinated the counterreconnaissance plan. Always did not want to take out all of the enemy's eyes too early, but he did want to identify where they were so that he could blind them when the time was right. This would consume many of the infantry resources needed elsewhere for the defense preparations. The aviation battalion would assist during the daylight hours the following day. The chemical officer suggested hitting likely enemy scout positions with a strike at a prearranged time. All agreed that this would diminish the chance of enemy reconnaissance escaping the sweep. The friendly scouts would be positioned in depth, with the triple mission of cutting off enemy penetration by reconnaissance elements, giving early warning of the main attack, and by continued reporting from all key terrain throughout the depth of the battlefield. Always would not fight this battle blindly.

Much had remained unsettled, but direct coordination the following day would allow further adjustments. A great deal of coordination was yet necessary on obstacle emplacement and the fire plan, to include the air force targets. The staff would have a full night's work ahead of them, both in planning and positioning the heavy equipment for the obstacles. Commanders would keep the work effort going through the night, but would be careful not to waste energy with inefficiencies created by the confusion that darkness brings. Before sunset the following night all movement rehearsals would be completed. Delta would conduct its rehearsals only after dark, and then one platoon at a time. Dilger did not want to give away his plan.

As the commanders left they had a good picture of what the defense would look like. The second meeting the next day, in which the overlay with all of the details of the order on it would be passed out, would take place at 1200. That would still be twenty-four hours before the expected attack. All of them, with the command encouragement of Lieutenant Colonel Always, were to get at least three hours' sleep during the night, and then to sleep at least three more hours during the next twenty-four hours. Similarly, they would ensure that all of their subordinates took the time to sleep, even if only a few men at a time. Sleep had become as important as ammunition.

The task organization seemed a little complex, but the commanders understood it and were already linking up all forces in the appropriate areas, orders having been passed by radio.

Team A
1 tank platoon
2 Bradley platoons
1 Bradley platoon (dropped off by Delta upon withdrawal)

Team B
2 Bradley platoons
1 tank platoon

Team C
1 Bradley platoon
2 tank platoons
1 ITV platoon

Team D
2 tank platoons
1 Bradley platoon (dropped off to A upon withdrawal)
1 ITV platoon

Team E
1 ITV platoon

1 infantry platoon (without Bradleys)
1 mortar platoon in direct support

Task Force Control
Scout platoon
Engineer platoon (reinforced with 5 operating bulldozers)
Air Defense platoon (4 Vulcan antiaircraft gun tracks)
Chemical Section (Smoke)

Supporting Elements
155 artillery battalion in direct support to Brigade, aviation battalion to be committed during the main battle, air force sorties (number uncertain) during main battle

Lieutenant Colonel Always and his jeep driver, Specialist Sharp, left the TOC shortly after 2100 to check on the troop work effort and to talk with the men working through the night. Working the opposite end of the battalion was Command Sergeant Major Hope. By 0200 Always pulled back into the TOC area, reassured that morale was good and operations were proceeding according to plan. He passed notes to the staff whose work he checked and modified where appropriate, and then retired, stretched out astride his Bradley in the open night air. He enjoyed the best sleep he had had in ages, awakening only after the sun had peeked over the horizon. As Always reached into his gear for his razor, Sharp handed him a cup of hot coffee.

It was still early when the battalion commander moved over to the TOC. Many of the staff principals were getting some sleep, but their assistants, most of whom had slept earlier during the night, were busily updating work reports on the defense efforts, coordinating fire plans and obstacle emplacements, and requesting of higher and supporting headquarters more of anything they could get—equipment, combat power, supplies, personnel, and so on. The assistant S-2, a senior sergeant, immediately reported to Always when the commander entered the TOC and updated him on the various bits and pieces of intelligence that had come in during the wee hours of the morning.

Command Sergeant Major Hope stopped by and joined Always in his second cup of coffee. They both agreed that this would be a day of maximum effort, but that they would insist on the highest degree of discipline throughout the battalion. Both were savvy enough to know that whenever pressure increased it was time to raise standards, not let them be lowered. As they set out upon their morning rounds, the two men—the ranking officer and the ranking noncommissioned officer—exuded an air of confidence and control. The effect on the men of the unit was electric.

In Bravo Company, young Sergeant Schwartz was bemoaning his attachment to an infantry company. He was the youngest tank commander in the company as well as the platoon, and his reputation over the preceding days of battle had gone sky-high. He felt that the tank, and particularly his tank, was the top gun in this style of desert warfare. He and his machine were as one; it responded to his touch like a well-trained horse. He resented an infantry commander pretending to have the wisdom to deploy his platoon. After all, what did the infantry know about tanks?

Bud Schwartz, though, was a professional soldier, and he knew how to take orders. He also knew how to give them, and he was now admonishing his crew to tighten the track tension on the left side. This was going to be a tough job, requiring the four of them to pull together to get it right. They had not gotten much sleep the night before, having to top off with fuel in the dark, reload ammunition that came up just before dawn, and prepare range cards at the crack of dawn. Then the platoon leader had arrived with the disconcerting news that the obstacle being emplaced to their front had shifted its positioning slightly, requiring the tank to be shifted as well, with a new range card to be computed along with the setting in of new target reference points.

Bud was a tough kid from Flint, Michigan. Tired and short of temper, he was trying not to take out his frustrations on his men. They would have to pull together as a team to get through the next couple of days. With the exception of his loader, Private First Class O'Donnel, who had joined them the previous evening, they were a well-drilled crew, having trained and fought together through all the battles. It was that new kid that bothered Sergeant Schwartz. He wondered if he would hold up in the heat of battle.

Always spent a good part of the morning with Captains Archer and Evans over on the right. He was troubled that the two of them essentially split a common piece of ground. Yet he felt that command at the gap was important enough to the plan to require a company commander to focus on the terrain immediately around it. It was while walking the sector that the colonel came across Sergeant Schwartz and his crew.

"How's it going, Sergeant? Men?"

"Just fine, sir. We'll be ready for them when they come." The tone was flat, but there was just enough irony in the sergeant's answer to make Always note the toughness in the man. He liked that, recognized the posturing of the delivery as big-city tough-guy talk, and took a second to eyeball the crew.

"How long have you been with us, PFC O'Donnel?"

"Since last night, sir." O'Donnel looked pale, self-conscious.

"Good. We need some fresh soldiers to give us the energy to put this together. You listen to your sergeant over the next few days. He'll tell you everything you need to know. Glad to have you with us." Always had picked up on the fear in his eyes.

"Thank you, sir." O'Donnel looked at his tank commander, whose stoic face showed absolutely no human emotion. He didn't know if he should be reassured or not. Always moved on.

An hour later he was with Echo Company, inspecting the progress on the obstacles being emplaced to prevent any infiltration down the sandy wash. Captain Evans was enjoying having infantry dismounts under his command. He was an infantryman to the core and, while content to command his antitank company, had missed the supervision of ground pounders.

He was talking to his commander. "Don't worry, sir. They won't get through here, not unless they commit everything they've got this way."

"We'll have to keep in mind that they just might. I'll take action if that happens. But you've got to buy me the time. Don't let them by without making them pay a helluva price."

And so it went throughout the morning and afternoon, throughout the operations order at noon, and into the evening. Commanders and soldiers talking and looking, checking and rechecking, offering suggestions, bouncing ideas off one another, reassuring each other, making adjustments, analyzing the terrain, shifting forces and obstacles slightly here and there, improving camouflage, and every so often taking an action intended only to throw any watching enemy off the track.

The counterreconnaissance battle was going well. Two BRDMs attempting to infiltrate wide around the flanks of the battalion were picked up by the scouts and killed. Four reconnaissance positions in the battalion's rear had been identified. Always committed a platoon-sized infantry patrol against the one he thought might be getting too much of an eyeful, but just marked and noted the other three. He would hit them with artillery just after dark, then put a little nonpersistent gas on their positions. They were on remote enough terrain to not endanger his own forces.

The toughest order to give had been the one to shift positions after dark. The men had worked hard all day. Instead of getting a respite that night, they would shift just enough to have to do it all over again. But Always felt it was necessary to give them just a slight edge over an enemy that had been plotting their positions all day, a chore made difficult for him by the numerous little deceptions played during the day. Now this one more would serve as further insurance that the enemy could not attack with a precise knowledge of the defensive plan.

The receptivity to the order had been enhanced, however, by the commanders' unanimous agreement that such a shift was necessary. They had explained the urgency of shifting position to their men during the day, and had allowed some of the preparatory work to be done discreetly during the daylight hours. The adjustments at night would not be that difficult.

Shortly after dark the three remaining identified enemy reconnaissance outposts were taken out. At the same time, possible outpost sites were also put under artillery fire. Under the cover of the noise, Captain Dilger rehearsed his movement back through the gap and into his secondary battle position at Battle Position 38. He also practiced dropping off his Bradley platoon with Team Alpha, the latter slipping into defensive positions prepared for them by Captain Archer's people. By 2300 all elements had rehearsed their movements, returned to BP 30, and set up for their initial defenses.

The enemy dismount probes began around midnight, slipping past Delta in BP 30 and honing in on Team Echo in BP 40. They were looking for an entrance into the sandy wash. Evans, however, had laid a number of antipersonnel mine fields in the first few hours after dark. It was in these that the enemy suffered his first casualties. Both Evans and Dilger had picked up his movement with their thermal sights, and had been talking to each other over secure radio for an hour before he closed. They watched as he worked his way through the mine field and stole closer to the obstacle blocking the way to the wash. Just a few hundred meters short of there, Evans put him under mortar fire. Nevertheless, he struggled on into the obstacle, only to be met there by a machine gun crew from one of the dismounted squads.

After a few minutes of chewing, he pulled back out, only to be put under mortar fire once again and finally eaten up by a Bradley positioned with Team Delta, which promptly shifted its position once it had done its work.

The report to Always made it seem like a complete victory. He did not suspect, nor did Evans and his dismounted infantrymen covering the obstacle, the hard-core determination of five of the enemy who remained through it all in the tank ditch in the obstacle complex. Scared and cramped, unable to move lest they give themselves away, they flattened their bodies against the bottom of the pit, waiting until times were a little more peaceful before pressing on with their mission.

Sergeant Schwartz cursed at the noise created by the ruckus over in Echo's area. He was trying to show O'Donnel how to fire the main gun. The kid seemed awfully dense, almost afraid of touching the mechanisms that would spit death out of the tube of the monster beneath their feet. The sergeant was tired and irritable. He could think of a thousand things he'd rather be doing than teaching his raw recruit how to handle an Abrams tank. But he defined it as part of his job and was determined to do it.

Always had been out with Captain Dilger, going over the plan once again. He knew that Delta was exposed and did not want to overlook anything in getting it back. Dilger appeared to have a firm grasp on things. There was no posturing by him to impress his commander, just a matter-of-fact discussion of the probabilities. He knew he was running a high-risk game, knew it was crucial to the battalion's battle plan, and knew he would have to shoulder the final decisions himself. Each man respected the other, not in least part for the responsibilities each was shouldering for the survival of both and for the survival of the battalion.

"You know that I'll be the last one out, sir?"

"Yes, I know. Don't wait too long to pull."

"No, sir. I won't."

They parted in the darkness, Dilger taking up his lonely post, waiting for the dawn and the attack he expected, Always heading back to the TOC, now located to the rear of BP 38, for a final update before moving to his battle position with Bravo on the ridge east of Hill 867. He had scouted for the best position earlier and decided that he could see the majority of the battle from this location, as well as bring in the helicopter battalion as it came up.

"Sir, higher headquarters reports that heavy forces are massing fifteen kilometers to our northwest." It was the S-2 speaking.

"That must be a report from Higher's long-range reconnaissance patrols. I hope they can tell us when they start moving." Always could picture the motorized rifle regiment massing its more than 150 vehicles, tanks, BMPs, antiaircraft guns, engineers, artillery pieces, BRDMs, personnel carriers, and command and control vehicles. For a second it gave him a cold chill. Then he shrugged it off and listened for whatever helpful details his intelligence officer could give him.

It was as he pulled into BP 32 and found Bravo's command post that the firefight over in Echo's obstacle broke out. Always went over to the artillery officer's track and listened to the calls for fire to the mortar platoon. He was impressed with the professionalism with which the incoming fires were adjusted. Captain Baker joined him inside the track for fifteen minutes, and with the artillery officer and the air liaison officer reviewed the plan should the main attack break in his sector.

With the report from Dilger that he had finished off the probing force, Always and his command group pulled up on the crest of the ridge, barely 200 meters from the leftmost extent of the tank obstacle cutting across Checkpoint 6. At 0100 he stretched out on the floor of his Bradley, with orders to be woken at 0315. Sleep came easily.

Captain Baker had also stretched out in his Bradley to get

some rest. At 0219 he was wakened by his radiotelephone operator. Captain Dilger was calling from Team Delta, reporting that one of his thermal sights had picked up dismounted troops moving toward his position. Baker walked over to his center platoon, talked directly to his platoon sergeant there in the platoon leader's Bradley, who in turn called over the wire to his platoon leader astride the obstacle. The lieutenant scanned the desert floor with his night vision goggles, picked up the squad running directly at him in a low crouch, waited until they were fifteen meters short of the obstacle, then blew his claymores lining the forward edge of the concertina wire. The entire squad fell dead and dying, legs severed, bodies shattered. The killing had taken less than a second. Always stirred in his sleep, then rolled over and drifted off again.

The enemy hiding in Echo's tank ditch flinched when the claymores went off over by Bravo. It had taken immense concentration for the five men not to move during the past two hours. Their sergeant was afraid the noise would awaken the machine gun team lying only twenty-five meters beyond the ditch. He had heard their heavy breathing begin only about fifteen minutes earlier. Now he would have to wait some more to make his move.

Captain Dilger was dozing off in the tank's cupola. He had put his men at 100 percent alert at 0130. He had gotten no rest prior to that, and so excused himself this little violation of his own policy as he drifted in and out of wakefulness. He was not depressed or gloomy, but somehow he was pessimistic about the battle ahead. It was not that he thought he would fail. He was confident enough about pulling off his mission successfully. It was just that he didn't feel he personally would make it through. There had been too many close calls in the previous days. He could not keep squeaking by. Sooner or later everybody's number has got to be up. He sensed his was coming.

Sergeant Schwartz had just gotten to sleep when Bravo blew

its claymores. He cursed out loud, looking over at the scared face of his loader sitting next to him in the cupola. He cursed again to himself and tried to get back to sleep, but it was a futile attempt. He was too pissed off.

At this moment Major Walters was bringing up the last load of spare parts to Charlie in BP 36. He had prioritized maintenance from front to rear during the last thirty-six hours, and right on schedule was bringing parts and mechanics together for final repairs. He would have everything fixed and in fighting condition by dawn. Of all the men in the battalion, he might have been the most tired. Only sheer stubbornness and pride kept him awake. At dawn he was going to move into the TOC. If the battle at the front went sour, he would take charge of it as it passed to the rear of the battalion. He calculated that with Charlie's tanks and those that made it back with Delta, he could crush any of the enemy that penetrated to his position. He hoped some would come. His blood lust was up.

At 0304 the enemy squad made its move. They slipped out of the tank ditch and slit the throats of the three men of Echo's machine gun team. They then returned to the obstacle and quietly started removing mines and cutting concertina wire. Two men worked at filling in the tank ditch with D-handle entrenching tools. None of Evans' men noticed. Many were asleep. Those who were awake figured the machine gun team by the obstacle had it well covered. Their attention was elsewhere.

The surrealism of the predawn hours shrouded the desert in a mystical panoply. Sleepy warriors in and astride mechanical beasts were caught between drowsiness and adrenaline bursts as they contemplated the fate that shortly awaited them. Hazy mists rose from the desert floor, ghostlike apparitions that folded over the equipment of war. Hundreds of men, flickering open their heavy eyelids, stared into the mysterious darkness as if to seek out the eyes of their intended killers staring back from across the endless wastes. Alternately, in an ironic juxtaposition of the savage and pacifist in each of them, they fondled a memento of more civilized times—a picture of a loved one, a letter, a locket—and the instruments of murder—a bullet, the edge of their bayonet, a hand grenade. Time was rushing toward destiny.

The obstacle in front of Echo had been unraveling for eleven minutes when Always awoke at 0315. For five more minutes he stretched and shook the cobwebs from his head, then choked down a tepid cup of coffee from a dusty thermos and unfolded his maps before him. At 0323 he raised Captain Dilger and discussed the wisdom of shifting 400 meters off position in the next half hour. The battalion commander was worried about the artillery barrage that would precede the attack. All other units had shifted after dark, as had Delta, but with the latter the chances were that the enemy had noted and recorded the change. It took four minutes for Dilger to agree that, although it would be a large pain in the ass, it should, and would, be done.

The stillness of the air gave Always his final opportunity to assuage the enemy's knowledge of his final defensive positions. At 0350 prepositioned smoke pots throughout the battalion's battle position gave rise to a series of clouds that covered companies and empty spaces in a sort of graceful shell game. Thermal sights cut through the smoke as easily as the darkness to look for the beginnings of enemy movement. The stage was set for the rise of the final curtain.

Six minutes after the long-range reconnaissance patrols reported the moving of the motorized rifle regiment from its assembly area, the horizon flared with light flashes—yellow and red, angry and intent. A split second behind the first flashes the sky beyond the horizon flickered with the reflections of yet more flashes. Then the first echelon of flashes sparkled again, and so back and forth in a wicked array of malign fireworks. Within minutes the ground around Hills 760 and 781 was being reshaped

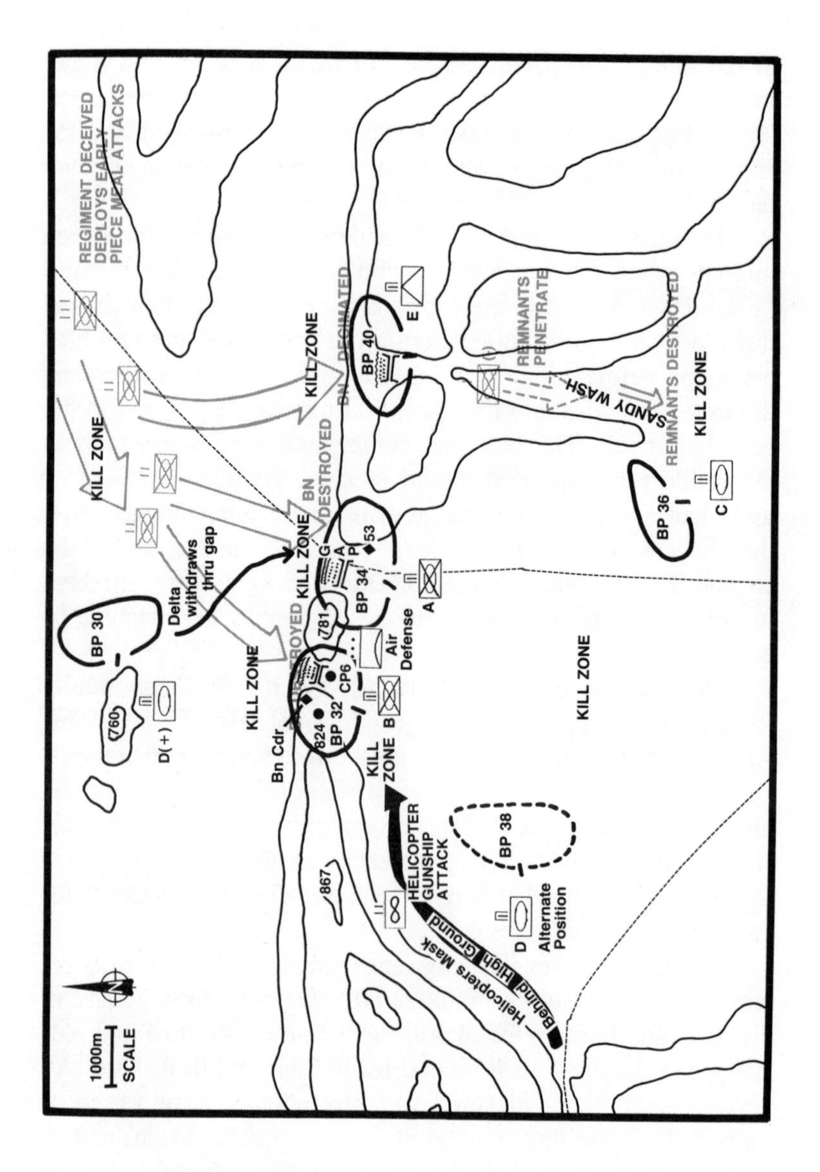
1000m
SCALE
N
REGIMENT DECEIVED
DEPLOYS EARLY
PIECE MEAL ATTACKS
KILL ZONE
KILL ZONE
BN DECIMATED
BP 40
E
REMNANTS PENETRATE
SANDY WASH
REMNANTS DESTROYED
KILL ZONE
KILL ZONE
BN DESTROYED
BP 36
C
53
G
A
P
BP 34
A
Delta
withdraws
thru gap
BP 30
KILL ZONE
DESTROYED
781
Air
Defense
CP6
824
BP 32
Bn Cdr
B
760
D(+)
KILL ZONE
KILL
ZONE
HELICOPTER
GUNSHIP
ATTACK
BP 38
867
D
Alternate
Position
Behind High Ground
Helicopters Mask

by giant explosions. It was the start of the enemy's artillery preparation. It was 0403.

The sound was terrifying. The earth shook and reverberated, protesting its violation. Men huddled in their holes, in their armored vehicles, against boulders, under their steel helmets, trying to pull their arms and legs closer together, unconsciously assuming fetal positions, faces twisted into grotesque masks of horror and fear. Shells rained from the sky, dozens falling every few seconds along short sectors of the front, reaching out, probing for hard metal and soft flesh. Hot shards of shrapnel ripped horribly through the air like giant, jagged scythes. There was no respite from the horror, the intensity of the barrage rising in crescendo with each passing moment. The front was alive with fire and steel, a blast furnace of destruction searing anything in its path.

For the most part the artillery barrage missed. The companies and platoons of the battalion were deafened by the clamor and pounded by the overpressures, but they had either moved far enough away from the artillery's target area or were dug in deeply enough, entire tanks now being sheltered by bulldozed holes, to withstand the terrible fire. Heads ached, nerves were frazzled, but with the exception of the unlucky, random victims, the defending soldiers survived the seemingly endless agony of the deadly deluge.

Battle Position 40 completely escaped the venom of the artillery. The enemy focused on 760 and 781, artillery pieces sighted on the last reported positions of Always' units. Under the cover and shock of the nearby churning, the tough enemy squad picked up its pace of reducing the obstacle. The enemy leader had two red-star clusters in his pocket as his only means of communication, his radio having been split by a 7.62 round in his initial dash to the tank ditch. He did not know if he would live to use them. It was the prearranged signal that the obstacle was open at the point of detonation. He did know that he would not live after he fired them. It was death either way. The thought spurred him on, a final effort to do the job right; it was his life's work now.

Captain Dilger was glad he had moved. His former position was taking a terrible pounding. As much as his eardrums ached, he listened intently for a change in the pattern, a slight lessening of the intensity. That would be the signal that the enemy was closing for the assault. He figured he would have five minutes to get in position before the onslaught hit. It would be a close race. He was glad he had rehearsed movement as much as he had. Seconds would count.

Major Rogers huddled beneath his Bradley. He had positioned himself close to the gap in order to ensure he was nearby whenever Dilger came back through. It put him right in the cauldron, but the bulldozer had done its job well. With the exception of a few nicks in the armor plating, the Bradley was relatively untouched, although the ferocity of the barrage was near maddening. Several times Rogers and his crew had had to fight off a sickening depression, an almost violent revulsion at the suffering they were undergoing.

Major Walters was waiting his turn at the TOC. He expected that the enemy would shift his fires deeper once he initiated his attack along the front. He hoped that the deception, night movement, and smoke cover would do the trick. Even if it did, things were going to be awfully violent in a few minutes.

Sergeant Schwartz was cursing again. He hated the artillery. It scared him, and he hated being scared, didn't want to show it, and so hardened himself with his cursing. O'Donnel was trying hard not to hyperventilate. The kid was shaking violently in the buttoned-up hatch. Unbeknownst to the two of them, a nearby hit had partially severed one of the left track pads.

The scouts were the first to sight the oncoming armor. The enemy was still in column formation when first sighted, about 6,000 meters out and closing fast. The attacking units were making about a kilometer every three minutes. In nine minutes they would be in range. The scouts called for and got artillery adjusted on them. Some armor careened to a stop or went up in a flash. The vast majority kept coming at no break in speed. It was 0424.

Dilger made his move. At a rush Team Delta fell in on the prepared positions in and around Hill 760, just as the enemy barrage lifted and concentrated around 781. TOW missile launchers popped up on every Bradley. Thermal sights scanned for targets. The sky had lightened appreciably, but the smoke hung heavily. In a few more minutes that too would be gone. For now the thermal sights were the best means of gunning down the enemy. Those who had them used them.

The ITVs out in front found the targets first. The first missile went out beyond its range, drifting off harmlessly. Then the platoon settled down and enemy BMPs and tanks began to erupt 3,000 meters short of their initial objective. They were deploying now on line, yet unable to see what was killing them, coming on hard nonetheless. Another volley of fire and the ITV platoon pulled out. They were the slowest of Dilger's vehicles and would need the most time to withdraw. One erupted in an explosion just as it began to withdraw. It was not clear what killed it.

The Bradleys opened up next, trying to pick the BMPs out of the crowd. It was clear by now that a reinforced battalion was coming at them. The enemy had begun to fire back in greater volume, surprised at the number of survivors of the artillery concentration. Dilger's FO (artillery forward observer) was giving the enemy hell, firing elongated target patterns along the attacker's route of march, catching his vehicles repeatedly, severing antennas, shattering gun sights and periscopes, occasionally immobilizing or destroying an entire war vehicle. The FO's timing was good as he estimated the speed of the advancing enemy and fired to his front. By the time the Bradleys withdrew, the enemy had lost an entire company.

It was now Dilger's tanks fighting it out, the enemy closing to 2,000 meters, the smoke clearing, the sky almost fully light. For five minutes it was a wild shoot-out, then the captain gave the order to pull back. He fired his own gun at two tanks, hit one, and pulled back in the trail of his company. As he moved toward the gap he came across one of his ITVs stuck in a wadi, its engine overheated. A Bradley had stopped to give it a battlefield tow, hastily set up, the final drives disengaged. Dilger pulled his tank around, called for a concentration of artillery, and opened fire on the nearest enemy; for a moment his attackers pulled into defile. It gave the Bradley time to put the ITV in motion, but the two connected vehicles could move only at a slow pace. Dilger held his ground.

A T-72 and its wingman were carefully working their way around to the company commander's left. Dilger could hear their engines above the roar of the battlefield, but he could not get a sight picture on them. A sagger missile sought him out, missing by inches. The captain pulled back his M1. "Move! Move! Damn it, move!" He was watching the Bradley-ITV crawl toward the gap. Another four minutes and they would have it.

Alpha Company had picked up the artillery mission, trying to cover Delta's withdrawal. The enemy was just out of range of direct fire. Rogers could see Dilger, could deduce just how the enemy was flanking him. He was praying for him to make it out of there.

As the crippled vehicle was towed through the gap, Dilger ordered his driver to move. The soldier responded, releasing the diesel fuel to pour over the engine, creating a dense smoke cover. But the winds were blowing it away as fast as it blossomed, and for brief, intermittent seconds Dilger's vehicle was visible to the closing enemy. An enemy lieutenant drew a bead on the M1, his finger poised to release his armor-piercing projectile. Dilger saw him just in time, slewed the main gun on line, and beat him to the punch. The T-72 flew apart and its occupants with it. A second later Dilger was dead, killed by the destroyed T-72's wingman, a sergeant who squeezed off his round even as the American captain was killing his lieutenant. The big M1 lay burning 600 meters short of the gap.

Rogers watched the black smoke rise into the air, swallowed hard, and ordered the FASCAM. In that instant the enemy launched his rush at the gap.

The remnants of a reinforced motorized rifle battalion came charging at the open area. Captain Archer's men opened up with everything they had. Sergeant Schwartz launched his first tank round, recoiled his M1 back deeper into its hole, reloaded, and pulled up to fire again. It took him six seconds to destroy his first two enemy vehicles. The defenders poured everything they had at their attackers. Tanks, Bradleys, missiles, artillery, machine-gun fire, small arms, and grenades hosed the enemy with molten lead. For seven minutes the battle raged in the few hundred meters surrounding the gap. Then the scatterable mines fell in and it was all over for the enemy. Sergeant Schwartz had destroyed eight vehicles. Not a single enemy had made it two hundred meters past the now-closed gap. Most of them were dead and burning on the enemy side of the mine field, the surviving crews desperately making their tortuous way away from the defenders. Artillery chewed them up as they withdrew.

The second battalion fell on Team Bravo. Its commander had seen the destruction of the lead battalion, realized the gap was closed, and staked his success on punching through Baker's people. The regiment's commander, forward where he could follow the battle, promised him fixed-wing aircraft to help drive home his attack. The radio message was intercepted, translated, and passed to Colonel Always at the same time it was relayed to the air defense warning net. As a result, when the two enemy jets winged in they were ambushed within seconds of passing the crest of Hill 781. They both crashed and burned before they could inflict any damage on the defenders.

It was too late for the enemy battalion to turn back. It bounced off Captain Baker just as Always committed the supporting helicopters, which approached from a masked position just south of Bravo's defensive battle position. The result was a duck shoot. The enemy was too concerned with the ground fight, then a raging inferno, to take the time to look up and see who was eating his lunch.

Always ordered Archer to shift his tanks over to Bravo to thicken Baker's defenses. It was on the way there that Sergeant Schwartz's track broke, leaving him stranded in the open just around the corner from Checkpoint 6. As he went out with his driver and gunner to check the extent of the damage, an incoming artillery round impacted a few feet from the tank. The driver and gunner were killed outright. Schwartz was flung against the side of the tank at the same time a huge chunk of shrapnel ricocheted off the armor plating and gouged across both his eyes. As he came to his senses, he realized he could not see.

O'Donnel rushed to help his tank commander, horrified at the sight of his two dead crew mates and repulsed by the red mash that had been his sergeant's face. Despite his own pain, Schwartz found himself trying to console his soldier.

"Okay, O'Donnel. It's okay now. Here, help me get back in the tank. This fight's not over yet." Schwartz was fighting to keep control over himself. He had not cursed since the artillery shell hit.

The two tankers climbed back in, the loader shifting to the tank commander's hatch under the direction of his sergeant, the sergeant taking up position as the loader. "Okay now, O'Donnel. Keep your eyes open. Remember what I told you about firing this thing. When the enemy appears, give me a yell so I know it's coming, then shoot him. I'll reload as soon as I hear the round fire. Got it?"

O'Donnel could not believe the toughness of his sergeant. He was going to load blind, a reflex conditioned by years of experience. He had not yet taken time to cover the wound across his eyes.

"Yes, Sergeant. I understand." His voice had sounded a little shrill. He swallowed, and fought to bring himself under control.

The tank had lost its track in an ideal location. A natural depression formed a miniature canyon that allowed the gun barrel to traverse along a narrow sector to its front, wide enough to track a vehicle for a few seconds but narrow enough so that the enemy would have to come directly at it to get a good shot. It was no accident that the tank had ended up there. Schwartz had picked out the route to Bravo Company on his reconnaissance. His foresight had payed off at the critical moment. If O'Donnel didn't lose his head, they had a chance.

The enemy was resolute. Despite the storm blowing apart his very fiber, he continued to press his attack. Soldiers dismounted in an attempt to reduce the obstacles in their path. They were eaten alive by the artillery and machine-gun fire. The enemy sent vehicles armed with grappling hooks to pull down the barbed wire. The first three vehicles were destroyed by direct fire hits. Nonetheless he continued to throw himself at Checkpoint 6. In the fourth try, a single-lane gap was opened through Bravo's defenses.

Baker and Always saw it at the same time. Instinctively the two warriors brought their own combat vehicles in line to shoot at the gap, but not before they gave the appropriate orders to bring other forces to bear. For Always that meant his close air support, which he had kept waiting five minutes out. The timing was tricky. He had to cut off his artillery and mortar fire along an air corridor so that he could bring them in safely. If there was too much of a time lag between cutting off the indirect fire and bringing in the aircraft, the enemy would gain a major advantage. Always brought the planes in a minute behind the shifting of fire. The enemy was caught by both systems.

For Baker the effort to plug the gap meant using every tank killing system he had. One of these was Schwartz and O'Donnel, who killed the first tank through the gap. It had appeared across O'Donnel's line of sight for four seconds. He needed only three to gun it down.

The climax of the fight at the Checkpoint took nine minutes. Only the first three were in doubt. After that it was merely a matter of the enemy being ground down to nothing. To his credit, he never took a step back. The aircraft got six. O'Donnel took out three more. Baker and Always got a tank apiece. The rest of Baker's people, organic and attached, did their part. Every BMP and T-72 was destroyed. Later on, Baker and his men would argue over who killed what. If all claims were honored, some 156 enemy vehicles were killed in the fight, a pretty good tally from an enemy battalion that offered only about forty targets. The argument was moot, however. The important thing was that the entire attacking battalion had been destroyed before getting out of Battle Position 32.

The third enemy battalion was heading toward BP 30 when the two red flares rocketed up from the ground. The obstacle denying entrance to the sandy wash had been reduced. In its final act, the enemy squad fired its flares and attacked into the nearest Echo Company position. They all died in the cross fire. But they had done their job: the way was open, if it was seized upon fast enough.

The regimental commander knew it was his only chance. He shifted his third battalion; moved into column with them; and under the cover of all of his remaining artillery, made a beeline for BP 40. Always' scouts picked up the change of direction first, but even as they called it in, the defending unit detected the shift. Always reached Major Walters by radio and directed him to pick up the rear battle, if it came to that. His XO was ahead of him, already having received the returning Delta Team and shifting them to reinforce Charlie's fires. He now moved himself to Battle Position 36. He almost hoped the enemy would get by Echo. He wanted a piece of them.

Captain Evans was outraged that the enemy had reduced his obstacle, but it was too late to undo that. All of his energy was focused on bringing his direct fire systems to bear on the onrushing enemy, the latter moving too fast to fire back. It was speed, not gunnery, the enemy commanders calculated, that counted now.

A company and a half were destroyed on the way through the gap. Evans suffered very little, aside from the ignominy of letting the enemy get past him. The embarrassment was redeemed a few minutes later, however, as Major Walters greeted the momentarily elated remnants of the enemy regiment with a reinforced company–sized ambush as they emerged astride BP 36. Only three enemy vehicles got by Walters and his people. They were policed up by the helicopter gunships that Always had shifted south to help out Walters.

The enemy regimental commander lay astride his BMP. He was a dying man, as much from the realization that he had lost his entire command as from the gaping hole in his armpit where once his arm had been. It was a bitter end. He did not know what had gone wrong.

Across the battlefield the acrid smells of gunpowder, burning diesel, and singed flesh mixed together to make a repulsive stench. Always looked with reddened eyes at the plumes of dark smoke rising everywhere. It occurred to him slowly that the fight was over, that he had won. He popped the hatch in his Bradley fully open and raised himself to a sitting position on the inside of the inverted cover. Major Rogers was already receiving the casualty reports. In a moment Always would make his rounds, check with his commanders, talk with his men. For now he wanted to savor his victory in silence.

CHAPTER 8

The Fruits of Victory

The final after-action review was finished. Lieutenant Colonel Drivon had been ecstatic, especially about taking credit for the great victory Always had achieved. Always had thanked him for his graciousness, then gotten away as quickly as he could to visit with his men one more time.

Elation was widespread throughout the task force. There is nothing as infectious as high morale, and nothing that gives rise to it quicker than victory.

Always felt an affection for each of his men. He tried to reach all of them, even though they were still busy reconstituting the force. Even after a momentous victory there is no rest for a unit. It still has to repair, reorganize, replace, and prepare for the next operation. All of these men had earned the right to leave Purgatory, but they would not go—could not go as professional soldiers—without first restoring a high state of order to their unit.

Despite their grime, despite the fatigue in their faces, there was no mistaking the joy and pride that showed through. It was in their eyes, carried in their voices, and sparkled from their very being. They had become a crack unit, and each man felt as if he had done it himself. In a way, each of them had.

Specialist Sharp epitomized the sense of accomplishment present in all the men of the battalion. He seemed to be even taller now; his walk had picked up an even greater jaunt. When he spoke to his fellow soldiers of "the Commander," it was with a certain reverence that at the same time belied his belief, his conviction, that he had helped to make him what he was. He indeed had, as had all of the other soldiers in the unit.

The mutual respect between the leaders and the led forged a bond that ensured that this particular battalion would be forevermore victorious. It had reached a state of invincibility. It might lose a battle, but it could never be defeated.

For the next three days the battalion worked on turning in its equipment. The ghouls at the issue point were merciless, but Always' men maintained their good humor. After all, they were on their way "home," and nothing could dampen their spirits. The battalion leaders stayed with the men, working alongside, trying to bring some order to the chaos that reigned over the turn-in. Their presence tended to solidify the strong bonds that had already grown between leaders and led. No more cohesive unit ever existed.

In the final hours of the last day Always took leave of his officers. He would be seeing them again shortly, but the time had come for him to make his way to his final reward. One by one he shook their hands, said his good-byes, and rendered a final salute. The look in the eyes of the company commanders revealed a deep respect for their commander, who returned it tenfold. Major Rogers and Major Walters, upbeat and optimistic as usual, gave a fond farewell and turned immediately to the organization of the departure of the battalion.

It was with a great sense of nostalgia that Always made the long drive back to Manix. He had had his success, but it did not diminish his sense of loss at leaving the men with whom he had been through so much, who were so much a part of him. Beneath the bridge where they had first met, Command Sergeant Major Hope prepared to say his last farewell to Lieutenant Colonel Always.

"How are you, Sergeant Major?"

"Well, sir, I'm doing just fine. I would say that our mission down here has been a success."

Always looked into the intelligent eyes of this fine noncommissioned officer. For a moment he felt terribly self-conscious.

"I know I've gotten my victory, but somehow that doesn't seem to be justification for my finally earning my way out of here. After all, it wasn't me who did it. It was you, and the rest of those guys, the whole bunch of them, from major down to private."

"Sir, there's an old saw in the army that a unit can be only as good as its leader. I don't know why it works that way, I can only say that it does. You gave them the freedom to be good. That's what counts."

Always pondered the words. "I'll miss these guys. Surely there could never again be such a fine collection of people. When the chips were down, each of those men did his part. It was more than a commander has a right to expect."

"Sir, you're right. Those are fine people. But every unit in our army has got fine people. In life you appreciated those in your specialized area, but you were limited in your exposure, and for that reason you developed a small bias that caused us to hold up your final reward for a bit. It was not your victory that earned you the way out of Purgatory as much as it was your understanding that it takes a lot more than a single arm, a single branch of the army, to give your men the wherewithal to win. More importantly, you have learned the worth of all of our people—the mechanics, the cooks, the medics, the artillery, the air defenders, the drivers, the infantrymen, the tankers, the engineers, the aviators, the air forces, and so on. In so doing you have lost most of your parochialism, and at the same time have become a better commander. That's quite an accomplishment."

"Well, Sergeant Major, I thank you for all your help. I know I couldn't have done it without you."

"You're very welcome, sir. It's always a pleasure to work with a good battalion commander."

"Will you be joining me?"

"In a little bit, sir. First I have a few things to check on with the men, then I'll be along—at least for a little while before another lieutenant colonel comes by here."

"I see. I guess I'm not the only small-minded SOB running around in our army. Well, I look forward to seeing you again. Thanks again, and Godspeed."

The two men shook hands, saluted, and parted company.

Always stood alone for a moment, watching the sergeant major drive the long trail back into the desert. He heard the cars racing by overhead on the highway, struck by the irony that the joy seekers in those speeding cars probably would never appreciate the pleasure he had just experienced, the pleasure of being with a crack battalion of American soldiers.

Then he turned and began his trek toward the setting sun. The sky never seemed so bright.

About the Author

James R. McDonough graduated from West Point in 1969 and currently serves as a U.S. Army colonel and director of the School of Advanced Military Studies (SAMS) at Fort Leavenworth, Kansas. He previously was posted to Supreme Headquarters, Allied Powers, Europe (SHAPE). McDonough has also commanded the 2d Battalion (Mechanized), 41st Infantry, part of the 2d Armored Division at Fort Hood, Texas, and served as a professor of political science at West Point, as an intelligence officer with Headquarters, U.S. Army Europe, and with infantry units in the United States, Europe, and Korea. (McDonough is in a unique position to write *Defense of Hill 781,* as he has gone through three rotations at the National Training Center and has been exposed to the hellish place on ten other occasions.)

McDonough is also the author of *Platoon Leader,* an account of his experiences as a young lieutenant assigned to the 173d Airborne Brigade in Vietnam, and *Limits of Glory,* a novel about the Battle of Waterloo.